Mulberry Cultivation

About the book

This book has been designed keeping in mind the requirements of the amateur mulberry cultivator and gives an indepth look the nitty-gritties of the industry. The various mulberry species have been enlisted with accounts of their descriptions, climatic requirements, soil conditions as well as their establishment, which includes land preparation, spacing, stakes and planting inter-cultivation, fertility, etc. Also information on pre and post-harvest requirements have been included.

Bearing a compendious amount of knowledge regarding the cultivation of mulberry, it is hoped that this book proves to be of immense worth to the reader interested in knowing more about this process.

About the Author

Dr. R.K. Patnaik, (b. 1963-) earned M.Sc. degree in Zoology from Berhampur University, Berhampur and Ph.D from the same University. Dr. Patraiis a popular science writer and has written 2 more books and 130 article in various Journals and Magazines. He is a member of several learned scientific societies.

Mulberry Cultivation

by

R.K. Patnaik

BIOTECH

2022

Biotech Books

ISBN10 81-7622-186-4
ISBN13 978-81-7622-186-3
EAN: 9788176221863

Published by : **BIOTECH BOOKS**
4762-63/23, Ansari Road, Darya Ganj,
New Delhi - 110002
Phones : 011 4300 3222
Email : biotechbooks@yahoo.co.in

PRINTED IN INDIA

Preface

With the burgeoning silk industry going places with its produce, it comes as no surprise that mulberry cultivation has acquired immense credential as an important sub-industry supporting the larger sector of sericulture. The mulberry leaf being the sole food for the silk bearing silkworms, mulberry cultivation is of great importance to sericulture, since quality and quantity of leaf produced per unit area have a direct bearing on the cocoon harvest.

This book has been designed keeping in mind the requirements of the amateur mulberry cultivator, and gives an in depth look into the nitty-gritties of the industry. The various mulberry species have been enlisted with accounts of their descriptions, climatic requirements, soil conditions as well as their establishment, which includes land preparation, spacing, stakes and planting inter-cultivation, fertility, etc. Also information on pre- and post-harvest requirements have been included.

Bearing a compendious amount of knowledge regarding the cultivation of mulberry, it is hoped that his book proves to be of immense worth to the reader interested in knowing more about this process.

Editor

Contents

1

Mulberry

Mulberry common name for a group of heterozygous, fast growing deciduous species of genus Morus of the family Moraceae. Mulberry is a multipurpose social tree of great economic importance. The leaf constitutes the chief source of food for silkworm (*Bombyx mori*); protein of mulberry leaves is the source for the silkworm to synthesize the two silk proteins, fribroin and sericin. The genus Morus is native to Indo-China and is widely distributed in the lower sub-Himalayan region up to the elevation of 2100 m. Mulberry plants are cultivated in the northern zone of Bangladesh, particularly in Rajshahi, where Bangladesh Sericulture Research and Training Institute has been established in 1962.

The genus Morus comprises of about 65 recognised species, of which *M. alba* is widely grown for rearing of silkworm. Mulberry plants are dioecious or monoecious. Inflorescence is catkin, and male catkins are usually larger than female ones. Flowers are small, sessile or shortly pedicillate, regular and cross-pollinated mainly by wind. Leaves are simple, alternate, entire, toothed, stipulate and petiolate. Mulberry plants are propagated by vegetative means. Locally cutting method of propagation is popular and is widely used but is restricted in a single season and usually done just prior to winter.

In Bangladesh Mulberry is not only associated with sericulture, but is also useful as fodder, timber, and medicine. The seeds contain 25-30% oil. Being a heavy leaf producing plant, it is also used as a shade tree.

Mulberry Cultivation

Mulberry (*Morus*) is a woody tree. Mulberry leaf protein is the only source for silkworm to biosynthesise the silk, which is made up of two proteins namely fibroin and sericin. It has been cultivated widely in India for sericulture.

Mulberry is cultivated in upland areas where water availability is less. Now a days drought and saline prone soils are also targeted for its cultivation. The success rate is very less due to poor rooting of the cuttings. This can be overcome by using ectomycorrhizal species at the nursery stage itself. Ectomycorrhizal species viz. *Pisolithus tinctorius* and *Laccaria laccata* were used at the Entomology Research Institute (ERI) to study their efficacy on the nursery seedling and establishment. *Pisolithus tinctorius* was found to be the best partner for mulberry in semi-arid conditions.

The inoculated plant stand was good and proved to be beneficial. The saplings even produced axillary buds, more number of leaves, increased leaf length, and increased biomass, which are absent or less in control plants. The root length and root number also increased due to ectomycorrhizal activity. This increased production can be attributed to the capacity of the ectomycorrhiza to translocate the required nutrients and minerals from soil to the plants and to produce large amount of auxin and cytokinins.

For inoculation, one gram of sorghum grain based inocula can be used per plastic bag/mulberry cutting. Application of up to 10 grams of rock phosphate/bag encourages the fungal establishment. Other fertiliser application can be deferred in the initial stages. This inoculation technology can be adopted in large-scale production of mulberry.

Use of Bio Fertilisers

Out of many micro organisms which are identified as bio fertilizers, Azospirillum, Phosphorus Solubilising Bacteria (PSB) and Vesicular Arbuscular Mycorrhizae (VAM) have a significant role in mulberry nutrition. Integrated use of these bio fertilizers will help in ensuring sustained productivity in mulberry.

Stimulation Growth

Azospirillum stimulates mulberry growth and increases the number, length and weight of roots besides bud development. It invades the internal parts of the root. Where it promotes root hair development and branching. It fixes the atmospheric nitrogen in the roots of mulberry and produces plant growth hormones.

Application of Azospirillum at 20 kg/ ha/ yr with well-decomposed farmyard manure in five split doses reduces nitrogenous fertilizer application by 25 per cent. Innoculation with Azospirillum also improves the economic characteristics of silkworm and silk. It improves larval characteristics (like larval weight, pupal weight, cocoon weight, shell weight and shell percentage) and silk characterestics (like silk filament length and weight of silk).

Use of Phosphorus

Phosphorus plays a major role in the balanced nutrition of plants to increase crop productivity. Only water soluble phosphorus is useful for the crop plants. The availability of phosphorus can be increased by applying phosphorus solubilising bacteria (PSB). It enhances the phosphorus use efficiency, produces the plant growth promoting substances, and also improves soil fertility and productivity.

The application of PSB at 10 kg/ha/yr with well-decomposed farmyard manure in five split doses reduces the quantity of phosphotic fertilizer by 25-30 per cent. Vesicular Arbuscular Mycorrhizae (VAM) is well known for their ability to absorb nutrients from the soil, particularly phosphorus. VAM plays a significant role in nutrient cycling in ecosystem. The external mycelium extends several centimetres from the root surface and it then passes the depletion zone surrounding the root and exploits soil microhabitats beyond the nutrient depleted area where the small rootlets or root hairs cannot thrive. Inoculation of VAM either individually or in combination with Azospirillum enhances the shoot and root length, number of leaves and the growth of plants.

Combined Inoculation

There is an increase in plant height, shoot biomass and leaf weight of mulberry due to combined inoculation of Azospirillum and VAM. The growth substances produced by Azospirillum are continuously released from root surface into the rhizosphere where the colony of Azospirillum grows with photosynthate supplied by the host plant. Inoculation of Azospirillum influences VAM infection of mulberry roots.

Biological Interactions

Mycorrhizal colonisation increases the Azospirillum chroococcum population in the rhizosphere and maintains it at a high level for a longer period of time. The beneficial effect of free-living nitrogen fixing organisms on plant growth is mainly due to hormone production in addition to nitrogen fixation. PSB survives for a longer period in the rhizosphere of mycorrhizal roots. The PSB renders more phosphorus soluble, while VAM enhances its uptake.

Today mulberry is cultivated in about 2.90 lakh hectares in India and the stress is upon productivity. Earlier the mulberry production per hectare was around an average of 12,000 to 15,000 kgs of leaf under irrigated conditions. Presently the average production has been tremendously increased due to the adoption of new technology like proper maintenance of mulberry garden and planting high yielding varieties while the average mulberry production per hectare has enhanced from 25,000 kg to 35,000 kgs.

Climate

Generally the Indian climatic conditions of all states are favourable to grow mulberry. The tropical temperature ranging from 240C. But at low temperature below 130C and above 380C the sprouting of buds and growth of mulberry will be effected. The favourable plantation season for mulberry is early monsoon. However, if abundant irrigation facilities are available mulberry can be planted at any time. The dormant season for growth of mulberry is November to March in temperate zones and rearing could be conducted only during may to October. In the tropical zones and rearing could be conducted only during May to October.

In the tropical zones this problem does not arise and mulberry grows all through out the year under favourable conditions. Always plantation season must be selected as per the above specified recommendations; too cold or too hot seasons badly effect the growth of mulberry. The mulberry can grow very well, with in the range of 600 mm to 2,500 mm. Mulberry growth will be considered optimum in elevated areas above 600-700 MSL.

Soil

Soil matters more in the growth rate of any plant. Mulberry grows in a wide range of soils. Mulberry is a deep rooted perennial plant. Therefore, the soil with good water holding capacity and good air penetration is more suitable. Soils ranging from clayey loam to loamy soils, slightly acidic soils are ideally suitable for mulberry growth, slightly acidic soils ranging the pH from 6.2 to 6.8 reveals healthy growth of mulberry plant.

Propagation of Mulberry

Mulberry can be propagated in two ways. i. Sexual, ii. Asexual

Sexual Propagation

In mulberry the sexual propagation is through seedlings, particularly the seed propagation carries a varied population, this to utilize in selection and hybridization. For seed germination certain prerequisites are needed to be fulfilled such as selection of quality seed, preparation of land, and the seed should be selected such that can definitely germinate. This is possible only when the seed is subjected to suitable environmental conditions, embryo of seed is alive, and healthy, in internal conditions of seed are favourable for germination.

The fresh seeds will have greater germination rate than the stored once. Seeds must be washed with fresh water until the flesh of fruit is with drawn and dried well, however minimum moisture percentage should be maintained, i.e. at least 6 %. Sowing of seeds may be by way of broadcasting or sowing in lines.

Asexual Propagation

In asexual type of propagation vegetative plant parts are used. In mulberry the propagation is mainly of three types.

— Propagation by Cutting.

— Propagation by Grafting.

— Propagation by Budding.

In case of mulberry, mostly through "Cuttings" only the propagation is practiced. By asexual method of propagation the inherited characteristics of parent stock can be retained. The desired characteristics can be carried to next generation.

The asexual propagation in mulberry is carried out by grafting, cutting, layering.

Cultivation and Practices

Mulberry can be cultivated in a wide range of soil. Favourable soils for mulberry growth are sandy loam, loam and clayey loam. The mulberry can be even grown on slopes of hilly areas, which are not prone to water logging. In case of slightly sloppy lands, proper drainage must be provided. No doubt the mulberry cultivation can be practiced in wide range of soils, however the leaf yield is the main factor. To increase the quality of leaf yield per hector all required favourable mulberry cultivation practices should be adopted, because the silkworm is fed by mulberry leaves only. The soil pH, water retaining capacity of field, quality of mulberry variety, temperature, environmental conditions fertilizer/manure application, irrigation practices and soil texture etc., all these factors very much influence on quantity and quality of leaf yield per acre per year.

Land should be prepared during early spring or late autumn. In Andhra Pradesh generally the most favourable season for plantation is pre-monsoon. The proposed land should be deeply ploughed during April-May. Since the mulberry is a deep rooted perennial plant and can last even for 10-15 years. The land should be ploughed repeatedly to loosen the soil and all gravel, stones and weed should be removed making the soil fine.

In irrigated land the recommended organic manure such as cattle dung compost @20 tonnes per hector or @ 10 tonnes per hectare in rainfed conditions must be applied and mixed with soil by ploughing 2-3 times. Land leveling is also important in case of plains. However, in hilly areas generally the land will be sloppy where pit system is adopted. In hilly areas June/July are the favorable where pit system is adopted in rainfed areas. In case of pit system the pits should be filled with loose soil and organic manure.

Fertilizers

Fertilizers also must be used as per the recommended dosage mainly N.P.K. in case of irrigated mulberry cultivation 120 kgs N,50 kgs of P, and 50 kgs of K per acre help for better leaf yield in Andhra Pradesh. Similarly the recommended dosage of N.P.K will be in the ratio of varied from place to place. Generally N.P.K. will be in the ratio of 2.5:1:1 for irrigated conditions and 2:1:1 of N.P.K for rain fed conditions. In hilly areas the recommended N.P.K. are 100:40:40.

Types of Mulberry

Mulberry variety suitable to that particular area of plantation specifically recommended. For hilly areas high yielding mulberry varieties, K-2, S-54 are recommended. And for the plain areas of Andhra Pradesh, Karnataka M-5 is recommended. The varieties are recommended on the basis of favourable response in the existing environmental conditions.

Spacing

Either 3′ x 3′ spacing is followed in pit system or 2′ x 2′ system is followed in row system in rainfed conditions, under irrigated conditions, the ridges and furrows are made at a distance of 0.30 to 0.45 m. Mulberry is planted at a distance of 0.10 to 0.15 m between the plants on either side of ridges along the row.

Inter Cultivation

To eliminate the weeds and to loosen the soil for good aeration, around mulberry plant, light hoeing should be done two months after plantation. This will facilitate the plant to absorb the water freely and plant growth will not be restricted by weeds by competing with mulberry plant for food, nutrition and water.

Pruning

Pruning of mulberry trees should be done after one year of plantation. By pruning the mulberry branches, leaf yield can be increased and production of leaf can be synchronized with silkworm rearing schedules all through the seasons. Pruning schedules controls the irregular growth of mulberry branches there by save wastage of nutrition and energy. Care should be taken during pruning, bark should not get peeled off since cut wounds do not heal, which leads infections and diseases.

Under pit system of mulberry plantation pruning can be carried once in June and again in November leaving a height of 8-10 cm above the ground level in wider spacing. Further under row systems pruning should be done at a height of 3 to 8 cm above the ground level in closer spacing.

Various methods of pruning are: - (a) fist form (b) Non-fist form.

(a) *Fist Form*: - Means cutting of mulberry plant each year at one place on main stem. The part gets thick & becomes fist shaped after few years therefore, the name. Easy control of mulberry diseases & pests possible in this method.

(b) *Non-fist form*: - Pruning done such that 2-3 branches allowed to grow from main stem (basal parts) & secondary branches also develop therefore, non-fist form. Resistance towards disease / pest control comparatively less in this method.

Leaf Harvesting

Along with leaf production, leaf utilization is also important. Leaf harvesting depends upon type of rearing practice. Leaf harvested by 1) leaf picking, 2) Branch cutting / shoot harvesting. Leaf harvesting in time is very essential as mature leaf contain low nutrient value & silk worms fed upon them become disease susceptible.

Leaf Picking

In India mostly leaf picking is practiced. However, this is more labour intensive, as availability of labour is not a big problem. Presently this method is being followed by most of the rearers. In a year 5-6 harvests are possible. First leaf harvest takes place after weeks of bottom pruning. The quality of leaves harvested with petiole is better. Where moisture content/nutrient value will not be dried up. Generally the farmer picks up the leaves well in advance and feed to silk worms. Leaves are harvested according to the age of silk worms. Tender leaves are picked up, chopped and fed to chawki worms. As the stage of silkworm advances mature leaves are fed to the worms.

Branch (Cutting) Feeding

Silk worms fed with mulberry branches hence name – branch feeding. This type of feeding adopted after 3rd moult of silk worm larvae.

Advantages of branch feeding:

a) Cost of labour minimized.

b) Quality (succulency) of leaves maintained (as leaves not detached from branches) for a longer duration.

c) Bed cleaning process easy & labour saving.

d) Easy to disinfect rearing equipment, if branch feeding adopted in shelf rearing / floor rearing.

e) Easy to preserve & maintain quality.

f) Easy to feed silk worms.

g) Wastage of leaf minimized

h) Hygienic conditions maintained effectively in rearing house.

i) Maximum utilization of mulberry leaves by silk worms.

Whole Shoot Harvest: This type of harvest reduces labour cost. Mulberry branches cut close to the ground level & fed to worms settled for 4th moult hence, uniformity in maturity of leaf observed. In this type, shoots harvested at an interval of 10 weeks & may extend to another 2-3 weeks depending upon availability of water and favourable climatic conditions. Around 5 harvests possible in this method.

Preservation of Leaf Transportation

Mulberry leaf should be harvested in the early hours of the day, as the day passes the quality of leaves in terms of moisture content comes down due to active transpiration. Hence the leaves are picked up during the morning hours.

Further, the leaves after harvesting should be transported in wet gunny cloth and preserved carefully, either in wet gunny cloth or leaf chamber made up of wood. Frequently water should be sprinkled on the leaf to keep the leaf fresh. However, leaf with water drops should not be fed to silk worms, which results in disease attack. To keep the nutritive value of leaves qualitatively fresh and for healthy growth of chawki worms always tender leaves (top leaves) should be picked up. As the worms mature, matured leaves should be fed. The leaves always must be succulent, healthy cocoon production and are susceptible to disease attack.

Always fresh leaves are recommended to feed the worms. Longer the preservation, lesser the moisture and the nutrition, hence temperature, humidity and moisture contents in the storage place should be 20°C–90% of relative humidity and cool, clean with frequent watering.

2

Black Mulberry (*Morus nigra*)

Morus nigra is a deciduous tree, slender but with numerous branches. Grows to 6-9 m in height, but it tends to be a bush if not trained when young. Leaves rough on upper surfaces and pubescent underneath, 7-12.5 cm long, often producing leaves of several different shapes, with 1 or more lobes, multilobed leaves often appearing on the same branches as lobeless ones; abnormally shaped leaves usually produced from stem shoots or sucker growths, and frequently by very vigorous young branches. Flowers held on short, green, pendulous, nondescript catkins that appear in the axils of the current season's growth and on spurs on older wood.

The flowers appear in 1.3 cm scaly clusters, female flowers ripening quickly into 1.3-2.5 cm blackberry-shaped edible fruits. Botanically, the fruit is not a berry but a collective fruit; the fleshy bases of pollinated flowers begin to swell and ultimately become completely altered in texture and colour, becoming succulent, fat and full of juice. In appearance, each tiny swollen flower roughly resembles the individual drupe of a blackberry. The colour of the fruit does not identify the mulberry species. It has been suggested that the generic name of the mulberry, Morus, was derived from the Latin 'mora' (delay), from the tardy expansion of the buds. An alternative explanation is that it comes from the Celtic 'mor' (black), referring to the colour of the fruit. The specific name refers to the dark colour of the fruit.

Cultivation

M. nigra is believed to have come originally from Persia but is now naturalized in parts of Europe that have a mild climate. It was known in the whole of southern Europe from the earliest times, and presumably introduced from Persia by the Greeks and Romans. It has been cultivated in England since the early part of the 16th century, probably having been introduced into Britain by the Romans.

During the reign of King James I, about 100 000 trees were planted in the eastern and midland counties of the UK, some still alive today. *M. nigra* was 1st grown for its wood and fruit and subsequently for rearing silkworm. Silkworm culture was 1st introduced to Italy, where *M. nigra* was employed for feeding the silkworm until about 1434. Then *M. alba* was introduced from the Levant and has been preferred for feeding silkworm ever since.

Natural Habitat

M. nigra trees are highly favoured in areas with long hot summers or extended droughts. A rugged species, *M. nigra* is fairly resistant to cold but grows best at lower altitudes when sheltered from wind and in coastal areas.

Geographic distribution

Native : Iran,

Exotic : Canada, China, Cuba, Dominican Republic, Germany, Greece, Haiti, India, Italy, Japan, Pakistan.

Puerto Rico, Spain, Sweden, Tanzania, United Kingdom, United States of America

Biophysical limits

Altitude: 0-2 000 m Soil type: *M. nigra* like a warm, well-drained soil, preferably a deep loam. Shallow soils such as those frequently found on chalk or gravel are not recommended.

Reproductive Biology

Trees are either dioecious or monoecious, and sometimes will change from one sex to the other. *M. nigra* trees do not begin to bear much fruit before 15 years of age. They are wind pollinated, and some cultivars will set fruit without any pollination, for example in California, USA. The self-fertile trees commonly produce 2 crops a year. *M. nigra* trees have been known to bear fruit for hundreds of years, and it is commonly said that the fruit of the oldest mulberry trees is the best.

Propagation and Management

Propagation Methods

M. nigra is somewhat difficult to propagate, as it tends to bleed heavily when cut. The best methods of propagation are by cuttings or layering and both are easy and successful. Grafts are also usually successful, although there may be incompatibility between white and black mulberries. Hardwood, softwood and root cuttings are also suitable methods. The tree can also be raised from seeds, which should be sown as soon as extracted from the fruit.

Tree Management

The tree can be grown in various forms: as a tall standard

with 1.8 m stem, as a low standard on a 0.9 m stem, as a bush, or as a pyramid-shaped tree. It can also be fan-trailed against a wall and is a good subject for pot cultivation. Standard trees require no pruning except for removing dead wood and thinning branches. Most forms, especially pyramid trees, need to have the lateral growth shortened to about 6 leaves to form spurs.

Wall trees need to have their branches trained 30 cm apart. It is not advisable to prune the trees heavily due to the plant's inclination to bleed—cuts of more than 5 cm in diameter generally do not heal and should be avoided. The bleeding will be less severe if the tree is pruned while it is dormant. Trees grow very rapidly, need full sun, adequate space (at least 4.5 m between each tree) and generally thrive with minimal fertilization.

Although somewhat drought resistant, they need to be watered in the dry season; if the roots become too dry, the fruit are likely to drop before being fully ripened. Pot trees need repotting each year before new growth commences. They are not easily killed, and careful pruning and cultivation have rejuvenated old specimens. Broken, undetached branches usually take root if they touch the ground.

Germplasm Management

Seed storage behaviour is orthodox; storage in airtight hermetic storage at -20 deg. C is recommended.

Functional Uses

Products

Food: The purple-black berries are large and juicy, with a good balance of sweetness and tartness that makes them the best-flavoured species of the genus. The ripe fruit contains about 9% sugar, with malic and citric acid. Berries can be eaten raw or dried, or used in pies, tarts, puddings, conserves, jams, or sweetened and pureed as a sauce; slightly unripe fruit is best for pies and tarts. Mulberries blend well with other fruit, especially pears and apples. The fruit is sometimes pounded to a fine powder and mixed with the flour for bread.

Fodder: Although inferior to those of *M. alba*, the leaves of *M. nigra* are also used for raising silkworms and have been used as a feed for domestic rabbits. Both cattle and goats browse the leaves and shoots; therefore, young saplings need protection.

Fuel: Pruned and dead branches are suitable for firewood.

Fibre: In Japan, a textile fibre is extracted from the bark.

Timber: The wood is a rich yellow that darkens over time to a rich golden brown. The wood is not affected by water and, because of its hardness, is used in joinery for articles subject to wear, for lathe work, and in the manufacture of barrels, caskets, snuffboxes and cups.

Tannin or dyestuff: The leaves yield a yellow dye, used mainly for heightening the sheen on silk. Alcohol: Fruits produce an alcoholic drink; for example, in Greece they are fermented for this purpose. In Devonshire, UK, they are sometimes mixed with cider during fermentation, giving the drink a pleasant taste and deep red colour.

Medicine: The main use of *M. nigra* in modern medicine is for the preparation of a syrup obtained from the ripe fruit employed to flavour or colour other medicines. It is a dark violet or purple liquid, with a faint odour and a refreshing, sweet-acid taste. *M. nigra* leaves are used in pharmacy for their astringent properties.

M. nigra has laxative and antipyretic properties. The bark is a reputed anthelmintic, used to expel tapeworms. A leaf, flower or root decoction can be gargled for diabetes; fever, sore throat and swollen vocal chords are treated with fruit juice. A parasitic fungus, known locally as 'meshimakobu', brown on the outside and yellow inside, grows on the old stems of mulberry trees on the island of Meshima, Japan, and is used medicinally there. Other products: A milky juice found in all parts of the plant coagulates into a kind of Indian rubber; it has been thought to give tenacity to the filament spun by the silkworm.

Services

Shade or shelter: The trees are fairly wind resistant with some cultivars being used as windbreaks. Ornamental: *M. nigra* is especially worth growing for its luxuriant foliage and picturesque form. Fruitless male trees are the ones most often planted near paved areas, as the fruit of female trees drops and can cause permanent stains. Pot cultivation for small indoor plants and bonsai trees is also popular. Boundary/barrier/support: When spaced correctly, the species can be planted as a live fence. Other services: Birds are so fond of the fruit that *M. nigra* is sometimes planted near more valuable fruit trees, such as cherries, to lure the birds away from the choice crop.

Pests and diseases

M. nigra is generally free of pests and diseases, although cankers and die-back can occur. A white powdery coating on the lower leaf surface is a symptom of mildew spread by the pathogens *Phyllactinia corylea* and *Uncinula geniculata*. It can be treated with sulphur dust, copper oxychloride, karathane and imazalil. In some areas, 'popcorn disease', in which the fruit swells to resemble popped corn, is an occasional problem. The disease carries over from one season to the next, so collecting and burning infected fruit help control it.

Edible Uses

Fruit

Fruit - raw, cooked or used in preserves. A delicious slightly acid flavour, it makes an excellent dessert fruit and can be eaten in quantity[K]. The fruit is juicy and refreshing, though it must be used as soon as it is ripe (from mid-August to September) otherwise it will start to rot. The fruit falls from the tree as soon as it is fully ripe. It is best, therefore, to grow the tree in short grass to cushion the fall of the fruit but to still make it possible to find and harvest[K]. The fruit can also be dried and ground into a powder. The fruit is up to 25mm in diameter.

Medicinal Uses

The mulberry has a long history of medicinal use in Chinese medicine, almost all parts of the plant are used in one way or another. The white mulberry (*M. alba*) is normally used, but this species has the same properties. Recent research has shown improvements in elephantiasis when treated with leaf extract injections and in tetanus following oral doses of the sap mixed with sugar. Analgesic, emollient, sedative.

The leaves are antibacterial, astringent, diaphoretic, hypoglycaemic, odontalgic and ophthalmic. They are taken internally in the treatment of colds, influenza, eye infections and nose-bleeds. The leaves are collected after the first frosts of autumn and can be used fresh but are generally dried. The stems are antirheumatic, diuretic, hypotensive and pectoral.

A tincture of the bark is used to relieve toothache. The branches are harvested in late spring or early summer and are dried for later use. The fruit has a tonic effect on kidney energy. It is used in the treatment of urinary incontinence, tinnitus, premature greying of the hair and constipation in the elderly. Its main use in herbal medicine is as a colouring and flavouring in other medicines.

The root bark is antitussive, diuretic, expectorant and hypotensive. It is used internally in the treatment of asthma, coughs, bronchitis, oedema, hypertension and diabetes. The roots are harvested in the winter and dried for later use. The bark is anthelmintic and purgative, it is used to expel tape worms. Extracts of the plant have antibacterial and fungicidal activity. A homeopathic remedy is made from the leaves. It is used n the treatment of diabetes.

Other Uses

A fibre used in weaving is obtained from the bark. A red-violet to dark purple dye is obtained from the fruit. A yellow-green dye is obtained from the leaves. Wood - used in joinery.

Prefers a warm moist but well-drained loamy soil in a sheltered sunny position. Prefers a light soil. Plants are very tolerant of atmospheric pollution. Trees are hardy as far north as southern Sweden. A slow growing but very ornamental tree, the mulberry is sometimes cultivated in gardens for its delicious edible fruit. The tree is not grown on a commercial scale because the fruit is too soft and easily damaged to allow it to be transported to market, and is therefore best eaten straight from the tree. There are some named varieties.

The mulberry takes many years to settle down and produce good crops of fruit, about 15 years being the norm. Trees fruit well in southern and south-western Britainbut they require the protection of a wall further north if the fruit is to ripen. This is a good tree for growing grapes into. It means that the grapes are difficult to pick, but they always seem to be healthier and free from fungal diseases.

Plants are late coming into leaf and also lose their leaves at the first autumn frosts though the tree in leaf casts quite a dense shade. Mulberries have brittle roots and so need to be handled with care when planting them out. Any pruning should only be carried out in the winter when the plant is fully dormant because mulberries bleed badly when cut. Ideally prune only badly placed branches and dead wood. Once considered to be a very long-lived tree, doubts are now being cast on this assumption, it is probably fairly short-lived. Plants in this genus are notably resistant to honey fungus.

The seed germinates best if given 2 - 3 months cold stratification. Sow the seed as soon as it is ripe if possible, otherwise in February in a cold frame. The seed usually germinates in the first spring, though it sometimes takes another 12 months. Prick out the seedlings into individual pots when they are large enough to handle and grow them on in the cold frame for their first winter. Plant them out in late spring or early summer after the last expected frosts. Cuttings of half-ripe wood, 7 - 10cm with a heel, July/August in a frame. Plant out in spring. A good percentage take, though they sometimes fail to thrive.

Cuttings of mature wood of the current season's growth, 25 - 30cm with a heel of 2 year old wood, autumn or early spring in a cold frame or a shady bed outside. Bury the cuttings to threequarters of their depth. It is said that cuttings of older wood up to 2.5 metres long can be readily made to strike. The cuttings are taken in February and planted 30cm deep in a shady sheltered position outdoors. The stem is wrapped in moss to prevent water loss by transpiration, with only the top few buds not being covered.

Cultivars

'Chelsea'

Long, dark mulberry-red fruit of fine flavour. Excellent eaten fresh, in pies or preserves, or made into win. A precocious tree, fruiting within 2-3 years of planting out.

'Kaester'

The large, elongated fruit is up to 4cm long and 12mm wide. Deep purple or black when fully ripe, the juice is bright red, the flavour very sweet with a good balance between sweetness and acidity. The tree bears heavily and is easily propagated by cuttings.

'Wellington'

The medium-size fruit is long, slender and cylindrical, it is about 30mm long and 10mm wide, the skin is reddish-black, the flesh soft, of good flavour. It ripens over a period of several weeks. A very productive tree.

Morphological and agronomical characterization of native black mulberry (*Morus nigra* L.): An Experimental Study in Sütçüler, Turkey

Mulberry is found from temperate to subtropical regions

of the Northern hemisphere to the tropics of the Southern hemisphere. There are three main species of mulberry: black (*M. nigra*), red (*M. rubra*) and white (*M. alba*). The black mulberry, which originates from Iran , is cultivated for its fruits in South Europe , Southwest Asia and is the most important species in the Mediterranean.

The black mulberry (Turkish name 'Kara Dut') is widely grown in Turkey. With its Mediterranean climatic conditions, Isparta Province in the Lakes Region is rich in genetic resources of native mulberry. In this region, Sütçüler district in particular has notable populations of black mulberry, which have been cultivated in gardens for their delicious edible fruits.

In addition to fresh consumption, another important use of mulberry fruit is for various confectionery products such as jam, marmalade, pulp and paste, and also for medicinal purposes, e.g. to treat mouth lesions. The fruits are gathered when ripe, sometimes sun dried on rooftops, and stored or kept in the deep freezer for use in winter.

In recent years, there have been studies on drying characters, pharmacological actions, flavour characteristics, quality characteristics and genetic diversity in different species of *Morus* L., especially *M. alba, Morus nigra,* which is used for its fruits rather than its leaves, has less variety than *M. alba*.

To the best of our knowledge, there have been no reports on the morphological and agronomic characters of black mulberry. The aim of this work was to select of the most promising germplasm from native populations of *M. nigra* L. in Sütçüler County for use in cultivation as well as for improvement programmes.

Material and Methods

Plant material and sampling

The study was conducted during 2000–2002 on six native *Morus nigra* L. populations from Sütçüler County in the Mediterranean region of Turkey. Sites of *M. nigra* populations were located in southern Isparta at about the same altitudes, i.e. 1063–1064 m asl. The research material consisted of seedling trees that had not been subjected to any cultivation practices such as irrigation, fertilization and pruning.

During the first year of trial, 153 genotypes from six locations were surveyed based on observations and personal communication with growers. Seventy-five genotypes that were healthy, free from any pests and had healthy growth were preselected. At harvest season, genotypes that produced less than 3 kg of fruit were discarded. In the second year, observations on tree, leaf and fruit characters were made on 10 selected *M. nigra* genotypes from four locations (S-1, S-2, S-3 from Karadutlar; S-4, S-5 from Sagrak; S-6, S-8, S-9 from Toptas and S-7, S-10 from Yaka). The phenotypic evaluation method is described below.

For evaluation of leaf traits, leaves were collected from 1-year-old shoots after harvesting of fruits. Twenty representative leaves were used for measurement of leaf characteristics. Sampling was from the middle part of branches. For characterization of fruit traits, fruits (about 2 kg per tree) were harvested during each trial year from different parts of trees, and immediately brought to the laboratory in ice-boxes. A first harvest was made, followed by one 15 days later. Thirty fruits from each tree were used for measuring fruit dimensions.

Phenological properties

The dates of bud burst, appearance of flowers, fruit colouring, beginning and end of harvest for each genotype were observed according to Lale.

Chemical analysis

Fruit juice was extracted with a kitchen blender and filtered. Soluble solid was determined with a digital refractometer. pH and total acidity were measured following AOAC methods. Total acidity was determined by titration with 0.1N NaOH to pH 8.1 and expressed as % malic acid.

The modified HPLC method according to Daood et al. and Alarcao-E-Silva et al. was used for vitamin C quantification. The analysis was carried out in a Shimadzu LC-VP HPLC system with LC-VP software, a pump (LC-6AD) and a UV–VIS detector (SPD-10A VP). An SGE column Wakosil II 5CI8RS 5 µm (250×4.6 mm i.d.) was used.

Chromatographic conditions were: flow rate 0.8 mL min –1; volume injected 20 µL; column temperature ambient; detection wavelength 254 nm; mobile phase deionized water (adjusted to pH 3 with H 3PO 4). Mulberry fruits (10 g) diluted with 10 mL 2% H 3PO 4 were homogenized (Ultra-Turrax) and filtered.

Of the elution solution, 1.5 mL (0.01M KH 2PO 4, pH 8.00) were added to 1.5 mL filtrate. Of this, 1 mL of solution was transferred from preconditioned Sep-pak C 18 cartridges (Waters) with 3 mL methanol and 10 mL H 2O and added to the column (Koyuncu et al. 2004). The cartridge was washed with 2 mL extraction solution. The eluates were mixed and used for HPLC separation.

Statistical analysis

Descriptive statistics were computed for quantitative characters. All statistical analyses were performed using the SPSS computer statistical program. The data were presented as the mean ± standard deviation.

Results and Discussion

All the black mulberry genotypes were healthy and free from pests. The bud burst dates of genotypes ranged from 29 March to 8 April 2001 . All genotypes first flowered at the beginning of April. The fruit colouring time varied from 15 to 21 May. Harvest began from 10 to 20 June and finished in September.

Genotypes had a range of 9–23 m for tree height, 7.30–17.20 m for tree crown width and 1.00–3.50 m for trunk circumferences. Some researchers report black mulberry trees up to 10 m with short trunks and widely spreading round crowns.

There were no differences for leaf shape: all genotypes had leaves that were unlobed, dentate and cordate. In general, leaf colour was dark-green with the exception of S-8 (green). Leaf-venation was observed as 'apparent' and 'slightly apparent'. Descriptions of leaf characteristics related to leaf shape and colour by Lale, Tutin, Verheij and Coronel and Yaltirik were in agreement with our results. According to these authors, black mulberry leaves were cordate at the base, usually undivided and dark green. The leaf length and width values were from 8.38 to 10.92 cm and 8.24 to 10.16 cm, respectively.

Genotype S-6 had the longest leaf petiole (2.63 cm) and genotype S-8 was found to have the shortest (1.89 cm). For leaf length, leaf width and leaf petiole length the results were similar to those found in the literature. For the fruit dimensions, genotypes studied had a range of 15.34–16.29 mm for fruit diameter, 22.33–25.15 mm for fruit length and 3.11–4.49 g average fruit weight. S-3 had the greatest fruit weight (4.49 g), followed by S-6 (4.36 g), S-5 (4.19 g), while S-2 had the least (3.11 g). In a study carried out in Izmir Province (Western Anatolia), fruit width, fruit length and fruit weight for black mulberry were reported as 1.56 cm, 2.70 cm, and 3.32 g.

For the physical characters of fruits, our genotypes were similar to cultivars 'Black Persian'(fruits >2.45 cm long and almost as wide) and 'Kaester' (originating from Los Angeles , with fruits. 3.8 cm and 1.27 cm wide). Gerasopoulos and Stavroulakis reported that the fruit weight of cv. Mavromournia mulberry varied between 4 g and 7 g depending on harvest stage. Mean fruit weights of S-3, S-4, S-5 and S-6 genotypes were heavier than 4 g. With regard to external colour, all genotypes have almost the same colour [L*a*b* (CIE)]. The lowest L* (the darkest colour) was obtained from genotype S-10 (L*=24.40) and S-1 genotype was dark red (a*=6.91).

The highest soluble solid was 18.36% (S-8) and the lowest was 13.91% (S-3). Lale and Özçagiran reported soluble solids of 15.5%, and Hulme 15%. Woodroof and Luh and Holland et al. reported water content of 85% while Güven and Basaran reported 14.20% in a study carried out in Çanakkale (eastern of Anatolia). The soluble solids observed in our study closely match those reported in literature.

The pH of fruit juice ranged between 3.35 (S-4) and 3.82 (S-6). The total acidity of genotypes studied was highest in S-4, and ranged between 1.24% and 1.64%. Lale and Özçagiran (1996) reported total acidity as 2.24% with pH 3.31 and Güven and Basaran as 1.48% and 3.70, respectively. Elmaci and Altug evaluated the flavour of black mulberry from the Aegean region of Turkey and recorded total acidity ranging between 1.51% and 1.79%, and pH values between 3.60 and 3.80. Thus, our results on pH and total acidity were similar to those reported earlier. Vitamin C contents were between 2.68 μg g –1 and 9.99 μg g –1 fresh weight; similar to values reported in the literature.

3

White Mulberry (*Morus alba*)

Morus alba is a tree which can reach over 20 m. in height but, in its cultivated form, is often coppiced, pollarded or pruned to a low-growing bush to facilitate the harvesting of fruit or leaves. The form of the tree can vary from pyramidal to drooping. The simple, alternate, stipulate, petiolate, light green leaves are cordate at their base but very variable in form, even on the same tree; some are unlobed while others may be almost palmate.

In temperate and sub-tropical climates mulberries are deciduous; under tropical upland conditions they are in leaf throughout the year. The tree generally has a deep tap root with little surface rooting which makes it suitable for use near crop land. Mulberry trees are commonly dioecious but may be monoecious, and sometimes will change from one sex to another. The flowers are held on short, green, pendulous, nondescript catkins that appear in the axils of the current season's growth and on spurs on older wood. They are wind pollinated and some cultivars will set fruit without any pollination.

Cross-pollination is not necessary. Male catkins are generally longer than female ones and are loosely arranged; after shedding pollen they dry and drop off. The juicy drupelets formed by the individual flowers on the catkin combine to form a sorosis, the characteristic mulberry fruit. Fruit colour is commonly black but can vary through light purple to white – the "white" in the plant's name refers to its leaf-buds. The juice of the fruit causes stains which are very difficult to remove from cloth.

Environmental Adaptation

Mulberries thrive in full sun and dislike crowded conditions; they prefer deep soils and need good drainage; they are frost resistant. Once these requirements are met mulberries thrive over a very wide range of warm temperate, sub-tropical and tropical conditions. Several species have been domesticated and innumerable cultivars selected over many centuries for both fruit and silkworm fodder so the plant's great range is not surprising. Care must, of course, be taken in finding the best cultivar for the site and purpose. Mulberries are found throughout the Old World sub-tropics and temperate areas with hot summers and have been widely distributed with the spread of silkworm-rearing.

Silkworm strains have been bred to fit the various availability of mulberry fodder; in areas with a short season the moths breed only once or twice yearly, where the season is long multi-voltine strains are used. For growing mulberries as livestock feed (as opposed to multi-purpose trees which yield some forage), however, areas with a long or continuous growing season would seem the most promising; this is reflected in the countries which report on such work in Sánchez 2002: Brazil, Italy, Cuba, Colombia, Costa Rica and Tanzania.

For India, Datta states that mulberries are found north of 280 and that they thrive with an annual rainfall of between 600 and 2500 mm. The altitude in the tropics is given as sea level to 1,000 m. Mulberries are, however grown at considerably higher altitudes in the Himalaya-Hindu Kush region. They are widely grown in Northern Pakistan and Afghanistan within irrigated areas and in villages.

In East Africa mulberry is grown as an occasional fruit tree up to at least 2 000 m. It also occurs on the Plateaux of Madagascar where there is a tradition of silk production, but *Bombyx* silk is probably a modern introduction; traditional Malagasy silk is made by several indigenous insects and a spider; the main moth is *Borocera madagascariensis* Boisduval which is fed on the leaves of the Euphorbiaceous "tapia" *Uapaca bojeri* Baill.

Its adoption has been expanded in Latin America, Central America and in the Caribbean, starting in the late 80s (Sánchez, 2000). Here, the commercial use of mulberry is first of all associated with regions where sericulture is common, and in other regions mulberries used to a lesser extent in orchards, recreational rural households and gardens.

Cultivation and Management

Propagation

Mulberries can be grown from seed or propagated vegetatively; they are also frequently grafted, especially fruiting cultivars. Methods vary from region to region, reflecting the plant's great ecological and genetic range. Mulberry is often associated with mycorrhizae including *Glomus mosseae* and *Glomus fasciculatum*. Katiyar et al. found, in southern India, that the amount of phosphatic fertilizer needed for mulberry cultivation could be reduced by 75 percent when plants were inoculated.

In China seedlings, cuttings, grafted plants and layers are all used. The propagation of improved F1 seedlings is recommended. Seed can be broadcast on a well-prepared nursery at 15 kilos per hectare; a straw mulch is used until the seedlings sprout two true leaves. The seed beds are weeded and thinned. About 180,000 – 200,000 seedlings per hectare are produced.

In South India three cuttings of 50 mm diameter and 25 cm long are planted directly at stake in 35 cm pits on a well prepared seed-bed leaving one bud exposed. Planting is done after the onset of the monsoon.

In the Indian Hills, however, direct planting of cuttings is not used, instead saplings of five to six months of age are planted during the monsoon.

In Japan mulberry propagation is generally carried out by grafting and by cuttings. Root grafting prevails because it is easy to handle and the grafted saplings have a high survival rate.

In Colombia recommends rooting cuttings(15 – 20 cm x 2 cm) in the nursery under plastic.

Installation

Deep well drained soils are suitable for mulberry cultivation, although in the wild it survives on dry stony sites. A pH near neutral is said to be optimal. A clean, well prepared seedbed is essential. High fertility is considered essential. Spacings for livestock production are still being worked out but are generally dense which would seem to indicate a high establishment cost.

In China mulberry is often planted scattered on odd pieces of land but in pure gardens the density varies by region. In Zheejiang (East China) 10,000 – 15,000 plants per hectare are planted and pruned to a trunk height of

0.5 – 0.8 m. In Guandong 90,000 – 120,000 plants are used and trained to a low, trunkless type. In some areas they are intercropped with cereals.

In India spacing varies with zone and conditions. In southern India both rainfed and irrigated mulberries are planted at 90 cm x 90 cm, in 35 x 35 cm pits. In temperate areas the spacing varies according to conditions (90 cm x 90 cm; 150 cm x 150 cm up to 300 cm x 300 cm) and pit size is increased to 45 cm x 45 cm.

In Japan mulberry gardens are usually planted with 6,000 – 10,000 plants per hectare. In densely planted fields, designed for mechanical harvesting, more than 25,000 plants per hectare may be used. Standard application of chemical fertilizer to mulberry gardens is 300 kg of N, 140 kg of phosphate and 120 kg of potassium per hectare for alluvial soil, and 300 kg of N, 160 kg of phosphate and 200 kg of potassium for volcanic ash soil. In either case, application of at least 15,000 kg of compost per hectare is recommended.

In Colombia Trujillo producing fodder for dairy stock, recommends between 25,000 and 32,000 plants per hectare with strict weeding. Three prunings are recommended to shape the plant; forage production increases up to the third year under a cutting regime. In Cuba Mártin et al. report 100 x 40 cm for fodder. In Colombia Boschini found that yield increased with planting density (in the range tested 60 cm x 60 cm –120 cm x 120 cm). In Brazil de Almeida and Fonseca used 300 cm x 60 cm.

In Japan the growth of mulberry is generally divided into three stages; new shoot development, growth and storage. New shoot development is a stage in which new shoots develop using reserves stored in stump or root the previous year. Growth stage is when carbohydrates and other substances are produced for vegetative growth by means of photosynthesis in leaves. Storage stage is when most photosynthetic substances are stored for next year's growth. In cultivated mulberry plants, however, photosynthetic organs are removed/harvested by pruning and leaf picking, thus disrupting the growth stage. The exploited mulberry plants resume growth using the remaining storage substances.

Mulberry fodder is generally cut and carried; the plant is not suited to continuous grazing. Optimal cutting intervals seem to vary with cultivars and probably ecological conditions. Weeding and maintenance fertilizer are generally necessary. There is no information on the longevity of stands grown for livestock but well maintained silkworm gardens are long-lived. In addition to the pruning used to form the initial bushes and the regular defoliation for forage, periodic pruning to maintain the desired bush type may be necessary; some are pruned on a short trunk, others are trunkless.

Seed Production

Mulberry produces masses of seeds and these are generally easy to harvest and germinate well. The seeds should be removed from the ripe fruit as soon as ready (for small quantities careful use of a liquidizer with plenty of water will separate seed from pulp). After harvest treatment of seed varies according to region, and probably with the different seasons of harvest and sowing. The seed can be dried and stored but immediate sowing is often recommended; stratification of *M. alba* seed in sand for several months is sometimes recommended.

Grazing Management

Mulberry is palatable to all classes of herbivorous

livestock and is well suited to such small animals as rabbits and cavies. It is not adapted to grazing. Most of the experimental large scale systems described in Sánchez (2002) are based on cut and carry.

Smallholder Systems

Mulberry is well suited for use as a fodder where it can be grown opportunistically around house-compounds, on spare pieces of land and along field edges, as it often is for silkworms; house-yard sites are often of high fertility. In this case spacing and training will depend on local conditions and much of the harvesting will depend on livestock needs.

High Pollarded trees are widely used in the western Himalaya and Afghanistan for multipurpose use – the tall (2.5 – 3 m.) trunk keeps the foliage above the grazing livestock. When poles are harvested they are usually left on the ground for a few hours and the small stock strip them of leaves and bark. In Pakistan, where silkworms are important in the mountainous tracts, up to 600 kilos of mulberry leaves are required to rear silkworms obtained from one packet of silk-seed. Thus, 22,000 to 24,000 tonnes of mulberry leaves are consumed each year.

Iqbal has reported that the value of the leaves required to rear one packet of silk seed is Rs. 200 to 300. The total value, therefore, is about Rs.10 million, based on an average price of Rs.250. Net return to the rearers is estimated at Rs.2 140 per packet, generating a total net income of Rs.45.6 million to the silkworm rearers in the country.

Conservation

Mulberry is generally used fresh although in the Western

Himalaya old dry fallen leaves are gathered as winter feed and litter.

Composition

Young mulberry leaves and twigs are a high quality fodder, readily accepted by all classes of stock.

Composition of Morus alba. White mulberry as % of dry matter

India	CP	CF	Ash	EE	NFE	Ca	P
Fresh leaves,	15.0	15.3	14.3	7.4	48.0	2.42	0.24

Crop Improvement

There is a very rich store of genetic resources for mulberry and, although it has been exploited over centuries for both fruit and silkworm feed, the study of mulberry specifically for the feeding of large livestock is just beginning. The many allied species are mentioned above.

"Generally, mulberry is a diploid plant with 28 chromosomes (2n=28). However, it is rich in ploidy and a lot of triploid varieties have been found especially among *Morus bombysis* Koidz. It is said that *M. cathayana* Hemsl. has tetraploid, pentaploid and hexaploid varieties. Both *M. serrata* Roxb., indigenous to India, and *M. tiliaefolia* Makino, originally from Japan and Korea, are known to be hexaploid. *M. boninensis* Koidz. is a tetraploid being endangered due to cross contamination with *M. acidosa* Griff. *M. nigra* L. is dexoploid (2n=308), the largest number of chromosomes among phanerogams."

Pests and Diseases

The crop is not greatly bothered by pests and diseases; the main ones are summarised by Sánchez as follows.

— *Maconellicoccus hirsutus* (mealy bug); causes Tukra in mulberry
— *Diaphania pulverulentalis* (leaf roller)
— *Spilarctia obliqua* (Bihar hairy caterpillar)
— Pseudaulacapis pentagona
— Naupactus spp.
— *Migdolus fryanus* Westwood (Longhorn beetle)
— "Cutting-Prone" ants (Sauba ant)
— *Meloidogyne* spp. (Nematodes)

Key disease of mulberry:

— *Leaf*: leaf spot, leaf rust, powdery mildew, leaf blight, bacterial blight, *Phylactinia corylea*
— *Soil borne*: root rot (white, violet), root knot, mulberry withering

4

Red Mulberry (*Morus rubra*)

Red mulberry *(Morus rubra),* called moral in Spanish, is widespread in Eastern United States. It is a rapid-growing tree of valleys, flood plains, and low moist hillsides. This species attains its largest size in the Ohio River Valley and reaches its highest elevation (600 m or 2,000 ft) in the southern Appalachian foothills. The wood is of little commercial importance. The tree's value is derived from its abundant fruits, which are eaten by people, birds, and small mammals.

Habitat

Native Range

Red mulberry extends from Massachusetts and southern Vermont west through the southern half of New York to extreme southern Ontario, southern Michigan, central Wisconsin and southeastern Minnesota; south to Iowa, southeastern Nebraska, central Kansas, western Oklahoma and central Texas; and east to southern Florida. It is also found in Bermuda.

Climate

Red mulberry grows under a variety of conditions. The

frost-free period ranges from 150 days in New England to 330 days in southern Florida. Total annual precipitation ranges from 1000 to 2000 mm (40 to 80 in). Best growth is in moist coves and flood plains in the southern half of its natural range. Mean annual snowfall ranges from zero in Florida to 150 cm (60 in) in New York.

Soils and Topography

Red mulberry grows on a variety of moist soils at elevations below 600 m (2,000 ft). Soil orders on which red mulberry is found include Alfisols, Inceptisols, Spodosols, and Ultisols. Seeds are carried great distances by birds so trees may be found on any soil that is not too dry. Best development is on well-drained, moist soils of sheltered coves along streams.

Associated Forest Cover

Associated species include sycamore *(Platanus occidentalis),* American elm *(Ulmus americana),* silver maple *(Acer saccharinum),* and sweetgum *(Liquidambar styraciflua)* in the southern parts of its range. Toward the north red mulberry is associated with American elm, red maple *(Acer rubrum),* boxelder *(Acer negundo),* and white ash *(Fraxinus americana).* It is a secondary species in succession and is seldom associated with primary invaders. Red mulberry is listed as a minor component in three bottom-land cover types: Cottonwood, Sweetgum-Yellow-poplar, and Sugarberry-American Elm-Green Ash.

Associated understory species are roughleaf dogwood *(Cornus drummondii),* flowering dogwood *(C. florida),* swamp-privet *(Forestiera acuminata),* Nuttall oak *Quercus nuttallii),* hawthorn *(Crataegus* spp.), and possumhaw *(Ilex decidua).* Herbaceous vegetation associated with red mulberry includes pokeweed *(Phytolacca americana),* stinging nettle *(Urtica dioica)*, poison-ivy *(Toxicodendron radicans)*, and greenbrier *(Smilax spp.)*. In the southern part of the range, red mulberry is often found in pastures and along borders of fields.

Life History

Reproduction and Early Growth

Flowering and fruiting

Red mulberry is dioecious but can be monoecious, with male and female flowers on different branches of the same plants. Both male and female flowers are stalked axillary pendulous catkins and appear in April and May. The blackberry-like fruit reaches full development from June to August. Each fruit is composed of many small drupelets which develop from separate female flowers ripening together.

Seed production and dissemination

Minimum seed-bearing age is usually about 10 years, but 1-year-old trees planted in an abandoned field in east Texas produced fruits at age 4. Optimum seedbearing age is 30 to 85 years; the maximum is 125 years. Good seed crops occur every 2 to 3 years. The average number of red mulberry fruits per kilogram is about 8,600 (3,900/lb); the average number of cleaned seeds per kilogram is 795,000 (360,000/lb). One hundred kilograms (220 lb) of fresh fruit yield 2 to 3 kg (4 to 7 lb) of cleaned seeds. Fruits that mature fall to the ground near the seed tree. However, because this relatively large, sweet fruit is a favorite food of most birds and some small animals, most of the fruits are eaten and dispersed by wildlife before they fully mature.

Seedling development

Seeds can be extracted from fresh fruits by mashing and soaking them in water, and then passing them through a macerator, where pulp and empty seeds are skimmed or floated off. Storage temperatures of -23° to -18° C (-10° to 0° F) are recommended for dry mulberry seeds.

Seeds can be sown in fall without stratification or in spring following 30 to 90 days of stratification at 1° to 5° C (33° to 41° F) in moist sand. In nursery practice, seeds are sown in drills at the rate of 160 to 260/m (50 to 80/ft) in rows 20 to 30 cm (8 to 12 in) apart. Germination, which is epigeal, usually is from 12 to 50 percent. One-year bare-rooted seedlings may be outplanted.

Vegetative reproduction

Red mulberry can be propagated from stem cuttings or by budding, but these methods are complex and require greenhouse facilities. The average rooting from stem cuttings taken in May, September, and January was only 7 percent, regardless of time of year. Red mulberry is a prolific root sprouter and can be reproduced by layering.

Stages to Maturity

Growth and Yield

Red mulberry is usually found as scattered individuals near streams or in other moist places. Stands of any size are not mentioned in the literature. Very little is known about the growth and development of this species. At maturity, red mulberry trees are an average of 5 to 21 in (15 to 70 ft) tall and as large as 76 in (30 in) in d.b.h., depending upon habitat conditions. In wooded areas, red mulberry is often an understory tree with a rounded, spreading crown.

Reaction to Competition

Red mulberry has been planted in the Midwest because its fruits are a valuable food for wildlife, but because it provides very little soil stability or cover for wildlife, it has not been planted widely. It grows best in open conditions but is classed as tolerant of shade as it often grows as an understory tree.

Damaging Agents

Red mulberry seems to be vanishing from at least a portion of its central range, possibly due to a bacterial disease. The effects and extent of this disease have not been investigated thoroughly, but it is known that red mulberry trees are becoming increasingly scarce.

The only noteworthy leaf pathogens of red mulberry reported in the United States are leaf spots caused by a species of *Cercospora, Mycosphaerella mori,* and *Pseudomonas mori.* Red mulberry also is susceptible to witches' broom, *Microstroma juglandis,* but the cause is unknown.

A variety of insects feed on red mulberry leaves, including the European fruit lecanium, *Parthenolecanium corni;* Comstock mealybug, *Pseudococcus comstocki;* and cottony maple scale, *Pulvinaria innumerabilis.* The American plum borer, *Euzophera semifuneralis,* and the mulberry borer, *Doraschema wildii,* attack twigs and stems of red mulberry.

Red mulberry has been rated as moderately tolerant of flooding as it usually withstands being inundated with up to a foot of water for a single growing season. It normally succumbs, however, after being flooded for two growing seasons.

Special Uses

The highest use of red mulberry is for its large, sweet fruits. These are a favored food of most birds and a number of small mammals including opossum, raccoon, fox squirrels, and gray squirrels. The fruits also are used in jellies, jams, pies, and drinks. In the past, the fruits were valued for fattening hogs and as poultry food. Red mulberry is used locally for fenceposts because the heartwood is relatively durable. Other uses of the wood include farm implements, cooperage, furniture, interior finish, and caskets.

Genetics

Red mulberry hybridizes frequently with white mulberry *(Morus alba)*, a native of China which has become naturalized throughout parts of the Eastern United States.

Edible Uses

Fruit - raw, cooked or made into preserves. Large and sweet with a good flavour, the fruit can be up to 3cm long. The fruit can also be dried and ground into a powder. This is used to make a delicious confection with almonds and other nuts. The fruit is soft and juicy, when fully ripe it falls from the tree and is easily squashed. Unripe fruits can cause stomach disorders. Young shoots and unfolding leaves - raw or cooked.

Medicinal Uses

The root bark is anthelmintic and cathartic. A tea made from the roots has been used in the treatment of weakness, difficult urination, dysentery, tapeworms and as a panacea. The sap is used in the treatment of ringworm. Another report says that the milky juice obtained from the axis of the leaf is used. The fruits are used to reduce fevers.

Other Uses

A cloth can be made from the fibrous bark. Wood - coarse-grained, light, very durable, not strong, soft, rather tough. It weighs about 45lb per cubic foot and is used for boats, fencing and cooperage.

5

Soil Fertility Management

Soil fertility for plants in reality is a very simple affair but difficult to explain. There are no miracle fertilisers, only plain and simple chemical nutrients, that are absolutely essential for plant growth. In moist sand loams plant nutrients are generally more accessible than in silt or clay soils although clay soils contain higher fertility resources. Nutrients became restricted or unavailable, in very wet or very dry soils for obvious physical reasons such as a lack of root absorption by most crop species.

Sixteen elements are essential for a plant to complete its healthy life cycle. These essential elements are (Where a plant gets its nutrients):

From Air and Water: Hydrogen, Oxygen and Carbon

From Soil and Fertilisers: a) Primary Nutrients – Nitrogen, Phosphorous and Potassium, b) Secondary Nutrients – Magnesium, Calcium and Sulphur, c) Micronutrients – Manganese, Copper, Chlorine, Molybdenum, Zinc, Iron and Boron. Soil testing is essential to be conducted for mulberry gardens at least once in two years to assess nutrient status, subsequently to add proper fertilisers in adequate quantities. Addition of organic manure can also be calculated based on fertility data. At the time of planting, it is very important to see that the cuttings are placed and the soil around is well compacted. Even the rooted plants obtained from nurseries may also be planted in this manner.

In fertile soils typical of the prairies, following repeated crop removal, nitrogen, sulphur and phosphate are the first to become growth limiting. In acidic soils below pH 5, typical of higher rainfall areas potassium, calcium and magnesium may also become growth limiting. When soil pH's are on the acid side (below pH 5) we add Ca in the form of lime, calcium hydroxide, limestone (calcium carbonate) or dolomitic limestone (a combination of calcium and magnesium carbonate), in tons per acre, to bring up the pH to 6 or higher.

Planting Methods of Mulberry

In any perennial crop, initial establishment plays an important role in subsequent growth and yield and mulberry is not an exception to this. Mulberry being a vegetatively propagated perennial crop, in commercial cultivation, to perpetuate the parental characters, the mulberry garden is generally established through stem cuttings. Suitable planting methods through cuttings have been evolved for rain-fed system of cultivation.

However, due to variation in soil moisture status, fluctuations in temperature and non-adoption of recommended methods of planting by farmers, the initial establishment is not satisfactory and variation in crop stand, large number of gaps, etc., are the common features seen in new gardens. As an alternate method, planting saplings instead of cuttings is very important from the point of above mentioned advantages.

Sapling

Sapling is a rooted cutting of specific age, that is, 100-120 days for low or high bush and about 240 days for small trees. Because of well developed root system, saplings get established quickly and grow vigorously.

Nursery Bed Preparation

A flat land nearer to water source is preferred as nursery site. Well drained land with loamy soil is ideal for nursery. The land must be ploughed or dug 30-40 cm deep and allowed for weathering in sun for 2-3 weeks. Land is again ploughed two or more times to bring the soil to fine tilth. Root stocks, pebbles and weeds should be removed at the time of ploughing and the land should be leveled. Divide the land into a number of small units to prepare the nursery beds. The size of each bed can be kept conveniently.

A bed size of 240 cm (L) x 120 cm (B) can accommodate 180 cuttings (row to row 20 cm and cutting to cutting in a row 8 cm distance) to raise 4 months old saplings. To raise 6 to 8 months old saplings, the same bed can accommodate only 100 cuttings (row to row 30 cm and cutting to cutting in a row 10 cm distance). Each bed on all sides should be separated by a bund of 25 to 30 cm width and height and provided with irrigation channel of 25 to 30 cm width and 15 to 20 cm depth. The following measures are required before planting cuttings in the nursery bed.

- Add 5 pans of organic manure - FYM/sericulture compost/ vermicompost to each bed and mix thoroughly with soil.
- In the case of clayey or black cotton soil, add 5 pans tank silt or sand per bed and mix with soil uniformly.

— In the case of red loamy or sandy loam soils, there is a possibility of termite infestation. As a preventive measure, 0.1% Chloropyriphos (5 ml per litre of water) can be sprayed to drench the soil of nursery beds (2-3 litres per bed).

Preparation of Cuttings

Plants selected (any popular mulberry variety) for preparation of cuttings must be healthy and free from scale insect, tukra infestation, etc. The shoots should be 6 to 9 months old and should have attained a thickness of 10 to 15 mm diameter. The stout lower portion and tender, green upper portion of the shoots are not fit to be used as cuttings. The middle portion of uniform thickness must be used. Cuttings for raising 3 to 4 months old saplings are prepared 15 to 20 cm long with 3 to 4 good buds. For raising 6 to 9 months old saplings, cuttings are made about 25 cm size. Sharp knife must be used to get clean cut ends without damaging the bark.

Cuttings are to be planted immediately after preparation. Due to unavoidable circumstances, if mulberry cuttings have to be transported over long distance or stored for 2-3 days, the same should be done carefully so that the cuttings do not dry. Keep the cuttings in wet gunny cloth, if they are not planted immediately after preparation. Dressed shoots are preferred for transportation more than whole shoots or stems as such, as the volume can be reduced considerably, than prepared cuttings, which dry too fast.

Dressed shoots are transported during cooler hours, especially during night over long distance. Prepared cuttings, if required to be stored, should be bundled with all buds in one direction and kept in wet sand bed with the buds pointing upwards under shade. It can be covered with a thin layer of hay. The sand bed and hay covering must be sprinkled with water daily to avoid drying or cuttings. Cuttings can be stored this way for 2 to 3 days.

Planting

Nursery beds should be adequately watered and made wet one or two days earlier to planting. Mark each row at a distance of 20 or 30 cm as required, with the help of a thread. In each row by using a pointed stick, make small hole at 8 or 10 cm distance, to insert cuttings. To avoid fungal attack, cuttings may be dipped in 0.2% Bavistin solution for 10 to 15 minutes. Cutting must be planted in the hole in a slightly slanting position, bud turned up, and exposing only one bud above the surface of the soil.

The soil around the cutting must be pressed firmly, after planting. In the case of poor rooting varieties, 200 ppm of NAA/IBA dissolved in 90% ethyl alcohol and diluted can be used to treat the cuttings before planting. The cuttings should be dipped in the solution for 6 hours and then used. Cuttings can also be inoculated with VA-mycorrhiza in the nursery bed to get better survivability and economisation of phosphatic fertiliser. For this 1.0 kg of soil based VAM inoculum/bed is required. The inoculum can be placed in furrows made for planting cuttings keeping the spacing mentioned above and the cutting can be planted on the inoculum and furrows be closed. By this technique, it is possible to save 75% of phosphatic fertiliser once the plantation is taken up with these inoculated saplings.

Irrigation

One time irrigation must be provided immediately after planting. Subsequently, irrigation is given once in 4 to 5 days in the case of sandy loam, and red soil once in 7 to 8 days in the case of clayey soil.

Weeding

Nursery beds must be kept free from weeds. At least two rounds of manual weeding are required, first after 25 to 30 days and second after 55 to 60 days of planting. Only light harrowing using weeding sickle or weeding fork must be attempted, taking care not to disturb the sprouted cuttings. Walking or sitting on the nursery bed should be avoided.

Application of Fertilisers

Chemical fertiliser must be applied in the nursery saplings attain 20 to 25cms height in about 55 to 60 days after planting, preferably after the second round of weeding. Following this, 500 g of ammonium sulphate or 250 g of urea can be applied per bed followed by light irrigation.

Plant Protection

The most common disease noticed in nursery is leaf spot. Application of 0.1% Bavistin twice at an interval of 15 to 20 days is recommended. Spraying of 0.1% Rogor to control thrips and 0.1% Metasystax to control mites are also recommended in case such infestation occurs.

Uprooting, Transporting and Planting of Saplings

Three to four months old saplings are ready for transplantation to establish bush system of plantation. For tree plantation or for raising high bush adopting wider spacing and high crown height, 5 to 6 months old saplings are used. Good saplings attain a height of 90 to 120 cm in 3 to 4 months and about 150 cm in 5 to 6 months in nursery bed.

Before uprooting of saplings from nursery bed, land preparation in the proposed plot should be adequately watered for easy removal. Nursery bed is loosened with the help of crowbar or guddali or pick axe to a depth of 30 cm and saplings are taken out one by one, without damaging the main roots.

After uprooting, long time preservation of saplings should be avoided as far as possible and must be taken to the planting site immediately. If long distance transportation is required, saplings must be packed in bundles, covered with wet gunny cloth or green leaves and taken to the planting site during cooler hours. Saplings must be planted immediately after uprooting and as soon as they are received at the planting site, in the pits prepared for regular plantation.

Kissan Nursery

Saplings ensure successful establishment of garden. It is, therefore, a viable proposition to raise saplings on large scale commercially and make them available to farmers. Saplings can be raised directly in flat/raised nursery beds. The production cost roughly works out to be Rs. 0.30 per sapling raised in nursery bed. At selling rate of Rs.0.50 per sapling, Kissan nursery is a quite profitable venture for the farmers. Plan of raising mulberry saplings through Kissan nursery on commercial scale is given in Figure 1.

Techniques of Raising Saplings

Low acreage of high yielding mulberry varieties in the field is mostly due to the non-availability of multiplication materials i.e., cuttings. At present, only 6 to 8 months old shoots are used for preparing cuttings. A technology has been developed to use the soft and semi-

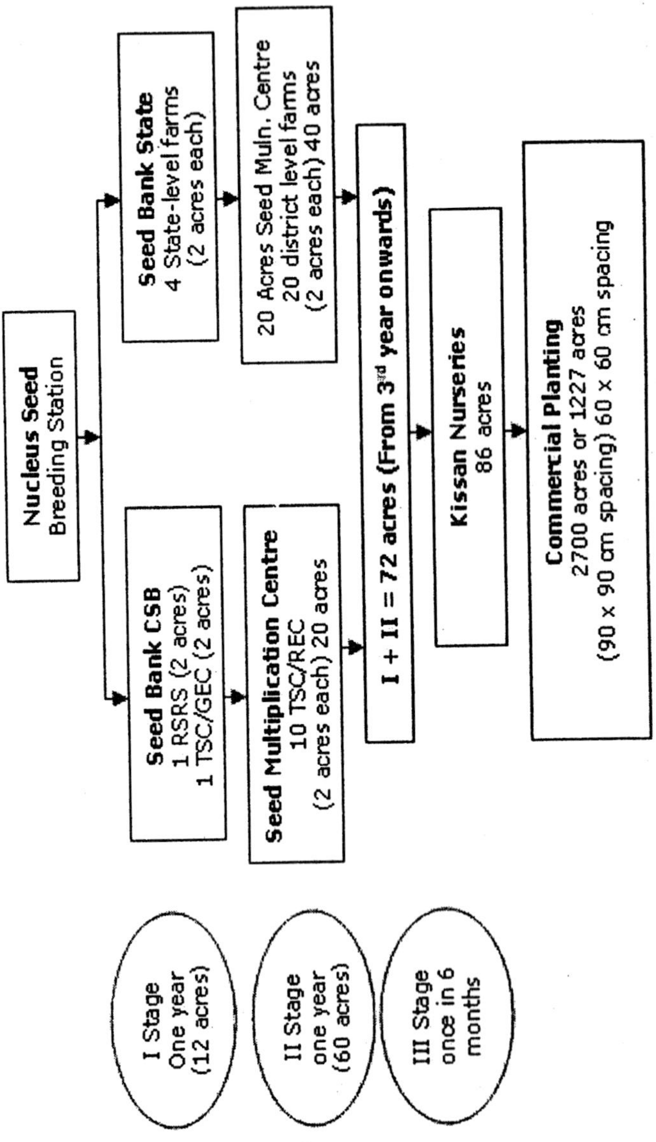

Figure 1. Seed multiplication programme for mulberry (to cover 2,700/1,227 acres of area)

soft wood green portion of the shoots as multiplication materials. However, soft wood (green) cuttings cannot be planted in the farmers' field directly or in the regular nursery. Saplings need to be raised in the nursery specially developed for this purpose and 3 to 4 months old saplings can be supplied to the farmers.

Preparation of Cuttings

Prepare 15 to 20 cm long cuttings with one or two top leaves intact. Cut the upper end of the cuttings horizontally above a bud and the bottom obliquely below a bud. Cuttings must be bundled (20 to 30 per bundle) with lower cuts arranged on the same plane. Dip the base of cuttings (2 to 3 cm) in 20 ppm NAA for 24 hours. After washing in water, plant them in the nursery bed.

Nursery Beds Preparation

Select a site with fertile soil and good irrigation facility. Sufficient compost or FYM must be mixed with the soil. Make raised beds of 5 cm high and 120 cm wide. The length of the bed may be kept at 240 cm, but can be made longer as per the requirement and convenience.

Plantation

Green wood cuttings prepared as above are planted in the nursery bed. It is advisable to plant in the evening to avoid bright sunshine. Cuttings are planted at 8 to 10 cm distance from each other in a row. About 5 cm of the base of cutting is inserted in the ground, shallower the insertion, better is rooting. After insertion, sprinkle water and cover the whole bed, as explained below.

Covering the Nursery Beds

Cover the whole bed with transparent polythene sheet supported with bamboo frames in a dome shaped manner to prevent the entry of outside air and to maintain high moisture inside the cover. The height of the cover at the central portion is kept at 50-60 cm above the bed. Above the polythene cover, gunny cloth is spread to serve as sunshade. The ideal temperature for rooting is 25-30°C with light intensity of 2000 lux. Care should be taken to maintain proper temperature and light intensity. If the beds are not prepared under the shade of tree, provide another layer of gunny cloth on the top.

Maintenance

Not so much watering is required. If the bed appears dry, irrigate the channel made outside the beds. By seepage, the nursery beds will be moistened.

Removal of Shade

When more than 10 roots per cutting appear, it is the right time to remove the shade. Generally, this condition comes after 35-40 days of planting. While removing the polythene sheet, remove one side one day and the other side on the second day. Fertiliser may be applied on the day of removing the sunshade and 2-3 weeks later. After 3 to 4 months, saplings are ready for plantation in regular plots. The methods are the same for uprooting, transport and planting, as in the case of raising saplings from hard wood cuttings.

Production Cost and Economic Viability

The technology is to facilitate fast multiplication of the limited availability of seed material of improved mulberry varieties. It also matches where mulberry is harvested by repeated shoot pruning and the plants do not attain the required maturity for raising saplings following conventional system. Commercially, it is possible to raise roughly 1.8 lakh saplings in 1 acre in a year costing Rs.0.40 per sapling.

Precautions

Do not transport green wood cuttings over long distance, as cuttings will have to be prepared with two green leaves intact, and in as fresh condition as possible.

Top Working Technique

Many high yielding mulberry varieties have been evolved recently. The time taken for the establishment of a new mulberry garden is minimum 6 to 8 months and roughly 2 years to come to the optimum leaf yield. The technique of bud grafting or top working can be employed to raise plantation of a high yielding variety (as scion) on the existing plants (stock).

Top working is a method of grafting a single vegetative bud with a piece of bark as scion, on the stock of existing plant. It is also called bud grafting. Bud grafting is the common vegetative propagation method followed in temperate regions of the country. This is practiced when the scion material is scarce and there are initial establishment problems like poor rooting.

The advantages of practicing this method is that, it helps the farmers not to skip too many crops, as the gestation period is comparatively very short. The establishment of the grafted material and the growth is quick enough to compensate the farmer with immediate mulberry crop portion in the existing mulberry garden, depending upon the availability of new variety. However, the limitations are high cost, long time, non-availability of skilled manpower and strict pruning care in subsequent years. Hence, needs precautions.

Procedure for Top Working

- A source for getting adequate number of buds of the new mulberry variety for grafting should be ensured. It is ideal to have a small separate plot of the new variety to ensure this.
- Existing plantation should be prepared to do top working, by cleaning the garden and removing the thinner branches. The stock branches that are 1.5 to 2 cm in diameter must be selected.
- Select the required, mature buds of the desired variety.
- Make a slanting cut with a sharp budding knife, below the bud and lift it upward to about 2-2.5 cm length above the bud and remove the bud gently.
- In a stock branch, select the internodal smooth portion of stem at height of 25 cm, only to a bark depth. Give another cut horizontally at the top of the vertical cut, which now appears as a 'T' mark.
- Gently lift the bark pieces on either side of the vertical cut for inserting the bud of the new variety, which is separated and kept ready.
- Insert the bud of the improved variety in the 'T' cut of the stock and tie the union tightly with polythene strip, exposing only the bud.
- After bud grafting, cut the stock branch a little above the grafted level.
- Irrigate the plot till sprouting of the grafted bud is ensured.
- Remove the sprouting shootlets of the old variety, regularly. Allow the growth of the grafted buds only.

Precautions

- — Basal pruning should be done always above the grafted level.
- — Shootlets of the stock (old variety) should be removed whenever such sprouting occurs.

Package of Practices

Mulberry can be cultivated both under soil moisture stress (semi-arid or rainfed) and irrigated conditions. Specific package of practices for mulberry cultivation has been developed aimed at improving both the leaf yield and quality under such stress conditions of rainfed regions. Package of practices mainly involve use of high yielding mulberry varieties like S-13 in red loamy soils and S-34 in black cotton soils, which are adapted to stress conditions. Measures for better establishment during initial stages of plantation and suitable agronomical practices emphasising soil moisture conservation and timely plant protection measures are essential.

Establishment of Mulberry Garden

Mulberry is a perennial plant. Once it is properly raised during the first year, it can come to full yielding capacity during the second year and last for 15 years in the field. There fore it is very important to establish the crop initially according to the scientific methods to obtained best results. To get better mulberry productivity, following package can be adopted:

Land Selection

Flat or slightly slopy land is suitable for mulberry cultivation. More slopy or steep lands need proper soil conservation measures through bench terracing, contour bunds and drains. Though mulberry can be grown practically in a wide range of soils, loamy to sandy loam and clayey loam soils are preferred. In the case of acidic (pH < 6.5) and alkaline (pH > 7.5) soils, necessary corrective measures should be followed through lime or gypsum application, respectively. Mulberry is a deep-rooted plant and the top soil should be sufficiently deep (60 to 90 cm).

Land Preparation

Planting should be undertaken during the commencement of South West monsoon. Therefore, land preparation should start in pre-monsoon period. Deep digging or ploughing of land to a depth of 30 to 35 cm is recommended, prior to the dry season, making use of the available soil moisture. The field is then left exposed during the summer, in order to kill weeds and soil-borne insects. During the pre-monsoon showers, the soil lumps must be broken and the land must be ploughed repeatedly to bring the soil to a fine tilth. The land should be leveled, and divided into plots of convenient size providing the contour bunds for proper harvesting of rain water.

Pit system of planting with wider spacing should be adopted for rainfed mulberry. Pits of the size 35 cm (L) x 35 cm (B) x 35 cm (depth) are dug at 90 x 90 cm spacing. Pits are then filled with a mixture of soil and FYM (2 kg of FYM/pit). One can use Sericultural compost or vermicompost in the pit to get better results.

Planting

Varieties S13 and S34 are recommended for rainfed conditions. Generally, mulberry cuttings are used for planting, but planting saplings is better. Branches of 8 to 10 months old and 10 to 12 mm diameter are selected for cutting preparation. Each cutting should be 20 to 22 cm length with minimum 3 to 4 healthy buds. The cut should be clean without damaging the bark or wood.

Greenish tender portion or too thick lower portion of the branches should not be considered for preparation of cuttings. Three cuttings are planted in each pit with a spacing of 15 cm between cuttings, in a triangle. Cuttings are placed deep, leaving just one bud on top, 2 to 3 cm above the ground and the soil around each cutting should be pressed well to keep the cuttings firm in the soil. It is risky to resort to direct planting of cuttings, under rainfed conditions. It is ideal to use 4 months old saplings (one sapling/pit) under such condition and support it with pot watering during initial period of establishment. Planting can be timed to receive adequate rainfall for 1 to 2 months period to the advantage for higher survival, quick establishment and maximum growth of plants.

Initial Care and Fertiliser Requirement

When the new garden is 2 to 3 months old after planting, inter-cultivation is done by weeding through light digging of the soil and applying the first dose of NPK @ 25 kg/ha through any complex fertiliser-Suphala/Vijay. Weeding is again done after another 3 months of first weeding, followed by application of second dose of nitrogen @ 25 kg/ha either in the form of urea or ammonium sulphate. This would enable the plants to grow well and establish fully.

Though the plantation may give enough foliage by the end of 8 to 10 months, it is advised not to take early leaf harvests before the plants attain full growth, which takes roughly one year under rainfed condition. The first pruning is done in the following rainy season, after one year. The plants are cut at a height of 8 to 10 cm above ground level, with a sharp pruning knife or saw without damaging the stem. Systematically cultivation practices must be followed from second year of planting.

Leaf Yield

By adopting the package of practices for the rainfed mulberry garden, it is possible to harvest 7 to 8 MT of leaf per hectare per year during the second year of establishment and 10-12 MT of leaf per hectare per year from third year onwards.

Azotobacter Biofertiliser

The high fertiliser requirement in mulberry cultivation in mulberry cultivation especially nitrogen (300 kg N/ha/year) is leading to increased cost of its cultivation. Fertilisers are not only in short supply, but also expensive and not available on time. Small and marginal farmers who cannot manure their garden in time in adequate quantity often harvest poor cocoon crop. Azotobacter biofertiliser, a cost effective supplement to chemical nitrogenous fertilisers in mulberry cultivation, is recommended to make sericulture more profitable. Besides, it reduces the deleterious effect of chemical fertilisers on soil health and also reduces the water pollution from nitrate contamination through leaching.

It is a bacterial preparation of live Azotobacter cells blended with suitable carrier material like lignite/peat in powdered from. It is capable of supplying nitrogen to the plants through biological nitrogen fixation and enhances plant growth. It is available in small polythene packets weighing generally 150-200 g and can be easily handled.

Recommendation

— Apply 20 kg Azotobacter biofertiliser/ha/year (to compensate 150 kg nitrogen) in 5 split doses @ 4 kg each time after every leaf harvest/pruning and intercultural operations.

- Use phosphorous and potash @ 120 kg/ha/year each in 2 split doses, as per recommendation for irrigated mulberry.
- Farmyard manure (FYM) should be applied @ 20 tonnes/ha/year as recommended for irrigated mulberry.
- Apply only 150 kg nitrogen./ha/year instead of 300 kg nitrogen in 5 equal split doses @ 30 kg each time after every leaf harvest/pruning and intercultural operations.

Uses of Biofertiliser

- Biofertiliser should be used before/after application of chemical fertiliser over a gap of at least 10-15 days.
- Before use, biofertiliser should be mixed thoroughly with dry powdered FYM (4 kg biofertiliser + 200 kg FYM each time) and applied in between the rows of mulberry making furrow.
- Furrows should be covered immediately after application.
- Irrigation should be given immediately after application of biofertliser.

Precaution

- Never mix chemical fertiliser with biofertiliser
- Keep biofertiliser away from heat and direct sunlight and use it before its expiry date.
- After application, land must be irrigated adequately and immediately to save live bacterial cells from desiccation.

— Collect biofertiliser from authentic sources to avoid spurious material.

Advantages

— Eco-friendly and reduces ill-effect of chemical fertilisers on soil health.

— Long lasting effect and sustainability.

— Cuts down cost by 50% in the total application of nitrogen fertiliser, thereby reducing, at least Rs.700/- per hectare per year, on fertiliser cost alone.

Preparation of Compost

Good quality farmyard manure is perhaps the most valuable organic matter applied to soil to enhance soil fertility and to increase micronutrients, which are essential for plant growth. Generally, in sericulture farms, the left over mulberry leaves from rearing bed and field and other waste including silkworm litter are not properly utilised in preparing compost of high nutrient value. The traditional method of preparation of compost is generally not effective. The loose heaps lie exposed to the sun, with the result the raw organic matter dries up quickly and does not decompose properly. Hence, it is essential to convert the sericultural farm wastes into valuable compost, adopting suitable technology.

The mixed farm wastes can be collected in pits of a convenient size. Pit size of 3 x 1 x 1 m is adequate to receive farm waste from 1 acre garden from roughly 2 harvests in 4 to 5 months time. Two such pits made side by side can be used alternately to compost all the farm waste from one acre mulberry garden. Each day's collection of various sericultural wastes like silkworm litter, left over mulberry leaf, weeds, etc. Should be spread in a thin layer sprinkled with a layer of fresh cow dung, ashes and water and compacted.

At the end of the rearing, the left over leaves in the garden, along with the young mulberry twigs can also be added to the pit. Some quantity of chemical fertiliser specially Superphosphate is also added to enrich the compost When the pit is filled and the bed stands 30 to 45 cm height above the ground level, it is then plastered with a 2.5 cm layer of a mixture of mud and cowdung. A thatched shed should be provided to protect the compost pit from rain and direct sunlight. To enhance the decomposition process, lignocellulose decomposing fungi like *Aspergillus* sp./*Trichoderma* sp./ *Belaromyces* sp. could also be added along with organic wastes.

Under above conditions, decomposition is anaerobic which is mainly a reduction process and high temperature is not generated. The plant remains, silkworm litter and cattle dung contain water, carbohydrates, organic acids, lignins and mineral constituents like phosphates, silicates, etc. During the process of composting, these are attacked by variety of microorganisms like bacteria, fungi, actinomycetes, protozoa and worms.

As a result of their activities, considerable portions of the residues, notably the carbon, nitrogen, phosphorous and potassium are rapidly liberated in available form for plant growth and well decomposed bulky organic matter is produced. The rate of decomposition depends upon the nature of the residues and upon the conditions under which decomposition takes place. During this process, organic nitrogen is reduced to organic acids and ammonia. The heat generated is sufficient to destroy all the harmful pathogens present in the material. Following conditions are required for satisfactory decomposition:

Nutrient content of various organic manures (on airdry basis)

Source	N (%)	P_2O_5 (%)	K_2O (%)
A. Green leaves			
1. Dhaincha (*Sesbania aculata*)	3.50	0.60	1.20
2. Sunhemp (*Crotalaria juncea*)	2.30	0.50	1.80
3. Wild indigo plant (*Indigofera tinctoria*)	1.80	0.20	0.60
4. Wild indigo leaf (*I. tinctoria*)	3.20	0.30	1.30
5. Rain tree leaf (*Glyricidia sepium*)	3.30	--	--
6. Red gram plant (*Cajanus cajan*)	2.80	0.40	2.00
7. Cassia siamea leaf or Ponnavarai (*C. auriculats*)	2.86	0.64	2.48
8. Pungam leaf (*Pongamia pinnata*)	3.31	0.44	2.39
9. Horse gram (*Volitions biflorus*)	2.71	0.53	2.21
B) Animal excreta			
1. Cattle	1.22	0.62	1.20
2. Goat	2.40	0.90	2.00
3. Sheep	1.93	1.30	2.30
4. Poultry	0.92	1.60	0.60
5. Horse	0.70	1.69	0.83
6. Silkworm litter	0.90	0.20	0.20
7. Vermi-compost	1.90	0.80	0.30
C) Various earths			
1. River silt	0.30	0.40	0.70
2. Tank silt	0.30	0.30	0.30
3. Press mud	1.22	2.20	0.42

— An adequate supply of available nitrogen in the form of organic wastes.

— Adequate moisture. Low moisture (10% and less) arrests the activities of most cellulose-decomposing organisms while high moisture content (80-95%) promotes anaerobic bacteria at the cost of fungi and most actinomycetes.

— A fine state of mechanical disintegration.

Chawki Rearing

Among the several factors that contribute to successful young age silkworm rearing, supply of highly nutritious mulberry leaves as feed is a vital one. The larval period and its maintenance is of utmost importance for the success of sericulture industry since it is the only feeding stage of the insect. The qualitative and quantitative requirements of the feed for silkworms differ at different stages of larval period. While it is generally established that the young age silkworms require mulberry leaf of higher succulency, moisture and nutrient contents, the late age worms feed on coarser leaf with less moisture content.

Other important factors determining the success of cocoons crop are the health, care and hygiene during the young age silkworm rearing-popularly called the "Chawki Rearing". Though importance of chawki rearing has been realised and several Chawki Rearing Centres (CRC) have been established to serve farmers, the problem of producing quality leaf in adequate quantity for chawki rearing centres has not been met successfully. To meet the suitable leaf requirement for exclusive Chawki centres any one of the training methods explained below may be adopted.

Harvests Schedule

Selection of Site and Land Preparation: Selection of site for establishment of the chawki mulberry garden is very important. The site must be more or less flat and in an elevated place, light textured, deep, sandy loam or clayey-loam soil. Site must be free from nematode and termite infestations and should not be low lying areas with excess soil moisture. The ideal spacing to be adopted is 60 x 60 cm.

Training Schedule: After an establishment period of about one year, the plants should be pruned at crown (20 cm above ground) during onset of monsoon. After 35 days of bottom pruning, harvesting of leaf is commenced for next 10 days (up to 2nd moult). At the end of rearing, the top terminal bud is to be clipped. Twentyfive days after top clipping, second leaf harvest as shootlets is to be done for rearing of worms up to second moult.

Thereafter, plants are again pruned at crown (80th day after 1st pruning). This cycle has to be repeated 4 times to get 8 crops in a year. Thus, plants are pruned at the crown (20 cm above ground level) 4 times in a year (after 2nd, 4th, 6th and 8th crops). Individual leaf picking is recommended for 1st, 3rd, 5th and 7th crops while the shootlets are to be harvested in 2nd, 4th, 6th and 8th crops for the purpose of rearing.

Manure and Fertiliser Schedule: The annual dose is 40 MT FYM/ha to be applied in two equal splits in 1st and 5th crops. The annual fertiliser dose is 225:150:150 NPK kg/ha in 8 equal splits @ 28:19:19 kg NPK/ha/crop. This can be applied as 127 kg Suphala 15 N: 15 P: 15 K plus 20 kg urea per crop.

Irrigation: 1.5 acre inch of irrigation water is to be provided once in 4-6 days by ridges-furrows method (34,000 gallons of water per acre each time)

Advantages

— Leaf yield: The above technology ensures a high yield of 29 MT of chawki leaves per hectare per year as against 7 MT/ha of chawki leaves obtained from common gardens. Almost 100% of the leaves produced are suitable for chawki rearing.

— Leaf quality: Qualitatively the leaves produced by this technology are superior (about 80% leaf moisture, and by dry weight 25% of leaf protein and 13% of leaf sugar) to leaves obtained from common garden (76% leaf moisture, and by dry weight 21% leaf protein and 11% leaf sugar).

Twelve Harvests Schedule per year

Establishment of mulberry garden on sound footing is a prerequisite for sustained yield of quality leaves per unit area over several years. Regarding this, the following activities must be carried out systematically.

Selection of Site: The site must be more or less flat and in an elevated place, light textured, deep, sandy loam or clay loam soil. Site must be free from nematode and termite infestations. It must have an assured source of irrigation.

Land Preparation: After selecting the site, the land must be given deep ploughing/digging. The weeds, root stocks and boulders must be removed. It must be allowed for weathering for about a month. The ideal soil pH range is 6.5 to 7.0. If the soil pH is above or below the desirable level, corrective measures may be adopted by applying required quantity if lime to acidic soil or gypsum and green manure to alkaline soil. The organic carbon content should be more than 1%; adequate quantity of organic matter should be incorporated in deficient soils. Entire land can be divided into smaller plots of convenient size with provision for irrigation channels and bunds.

Spacing: Spacing for planting the saplings should be 60 cm between the rows and 60 cm between the plants in a row (60 x 60 cm). With this spacing, about 10,000 saplings can be planted in an acre (0.4 ha.) of land.

Planting Material: Healthy, 100-120 days old saplings of 80-90 cm height can be used for planting. Because of already developed root system in saplings, plants get established quickly and grow vigorously.

Planting: Planting work can be taken up soon after the onset of regular South West monsoon. One sapling per pit be placed deep and straight in the pit/trench by removing the soil. After this, the soil around the sapling must be pressed firmly. When planting work is over, adequate irrigation has to be given immediately and in the event of failure of rain.

Intercultivation and Maintenance During Establishment: About 1½ months after plantation, alight weeding/hoeing must be done. Second weeding must be done around 3 months of planting. Weeding operations must be thorough and should be regularly carried out. The plantation taken up during the monsoon period will have the advantage of receiving fairly distributed rain from June to October. If there is no rain or not adequate for a period of over 10 days, supplemental irrigation has to be given.

After the cessation of the monsoon, irrigation has to be given regularly at an interval of 6-7 days, but this frequency varies depending on weather conditions and types of soil. About 1.5-2.0 acre inches (34,000-45,000 gallons) of water is required per irrigation. Irrigation can be given by furrow system and whenever possible, drip or sprinkler system can be adopted. Mulberry grows vigorously under assured irrigation and soil fertility. First dose of chemical fertiliser should be given as booster dose at the rate of 50:50:50 kg NPK/ha after second weeding and 3 months of planting.

By sixth month, plants would be ready for first harvest of leaves. After the first light harvest of leaves followed by weeding, a second dose of chemical fertiliser must be applied at the rate of 50 kg/ha. In the interest of good establishment of the plantation, it is not advisable to go for regular leaf harvest till the completion of one year of planting. After 6 months of planting, 3 light leaf harvests can be made till basal pruning is taken up. In the event of occurrence of fungal diseases, spraying of 0.1% Bavistin and in case of attack of insect pests, spraying of 0.1% Metasystox has to be done. Care should be taken to allow required safety period before harvesting leaf for brushing.

Management of Established Garden and Leaf Harvest: About one year of planting, with good maintenance, plantation will be established well. Following the good establishment, regular management and leaf harvests can be initiated. The aim of developing exclusive chawki mulberry garden is to get suitable nutritious leaf for chawki rearing continuously. In this direction, the following cultural operations must be carried out systematically.

— *Pruning*: The first basal pruning must be at 30 cm above the ground, preferably during the onset of South West monsoon. Following the first basal pruning and 3 leaf harvests, at an interval of 15 days, first middle pruning at 60-70 cm above the ground has to be done. Following this, 3 more leaf harvests can be made. Again the same sequences will follow. Totally 2 basal pruning and 2 middle cuts must be done in a year. After about 20 days of each pruning, weak branches can be removed.

— *Weeding and digging*: After thorough weeding, digging must be done to loosen the soil, to stimulate the plant growth and to facilitate the application of manures. Following digging operation, ridges and furrows should be formed.

— *Manuring*: Yearly twice, bulk organic manure in the form of farmyard manure or compost should be applied at the rate of 15 tonnes/ha/time and incorporated into the soil. This may be carried out within 15 days of each basal pruning. Thereafter (8-10 days), application of chemical fertilisers should follow. The total annual requirement of chemical fertilisers is 240: 140: 140 kg NPK/ha. The fertilisers should be applied in 4 equal split doses corresponding with 2 basal and 2 middle pruning.

— *Irrigation*: In light sandy loam soils, more frequent irrigation at 6-7 days' interval may be found necessary, while in heavy clay loam soils, irrigation at 8-10 days' interval may be found adequate. It is also important that at the above intervals of irrigation, adequate quantum of water should be supplied as irrigation at a time. It has been estimated that 1.5 to 2 acre inches of water is required per irrigation.

— *Plant protection*: To improve the leaf quality as well as the productivity, whenever necessary, plant protection measures must be followed. But, these measures can be taken up only after pruning and 15 to 18 days before leaf harvest for brushing.

— *Leaf harvesting and preservation*: From brushing to two feedings after second instar, the silkworms are fed with young leaves. The leaves to be harvested are from below the largest glossy leaf, which is yellowish green in colour. The cardinal point is shoot tip and it should not be removed during any crop. Below the glossy leaf, about 3 leaves during the first (1-3) and about 3 leaves (4-6) during the second instar can be harvested. Usually in cool hours (morning and evening) harvesting has to be done. While harvesting, care must be taken not to damage the shoots. Due attention must be given to harvest required quantity of leaves in the shortest time possible from the plot and to bring that to rearing house immediately. Excess leaf harvest and longer preservation must be avoided.

Silkworms grow best when fed with fresh mulberry leaves, which are rich in nutrients and moisture. Under tropical conditions, driage of leaf is faster. Usually, the leaves are harvested twice a day and are preserved for successive feedings, depending on the necessity. During this period, the leaves should be preserved in cool and clean places in order to preserve their succulency. It is batter to preserve leaves in a leaf chamber or bamboo basket which is lined with gunny cloth, that can be kept wet by sprinkling water at frequent intervals.

Harvested leaves must be sprinkled with water in summer season, if necessary. Alternately, chawki mulberry leaves can be stored in a mud pot, which is placed in the moist sand and mouth of the pot must be covered with wet cloth. Also, chawki mulberry leaves can be stored on a flat moist sand bed covered with wet and clean white cloth. Again after arranging the leaves on the sand bed, they must be covered with wet white cloth.

Economics: To work out the economics of chawki leaf production, three models of 0.2 ha (yearly 5,000 kg leaf and 30,000 Dfls), 0.4 ha (yearly 10,000 kg leaf and 60,000 Dfls) and 0.8 ha (yearly 20,000 leaf and 1,20,000 Dfls) have been considered. This consideration is based on availability of 10,000 kg leaf/0.4 ha (1 acre)/year by 12 crops. Also, as 15 kg leaf requirement/100 Dfls. The cost of chawki leaf production works out to be Rs 2.60 per kg, with depreciation and Rs.1.77 per kg without depreciation.

Leaf yield: This method will ensure roughly 25 MT of chawki leaf per hectare per year in 12 harvests. All the leaves produced are fit for chawki rearing, unlike in regular mulberry garden where only about 20% of the total leaf yields is fit for chawki rearing, which is about 7 MT per year per hectare. Leaf yield in exclusive chawki garden raised is adequate to rear 1,20,000 layings per year per hectare.

Leaf Quality: Qualitatively the leaves produced in chawki garden are far superior with 75 to 80% moisture content, and by dry weight basis 25% leaf protein and 13% leaf sugar to leaves obtained in general commercial garden, with 70 to 74% moisture content, 20 to 21 leaf proteins and 11% leaf sugar.

6

Morphology of Mulberry Plant

Silk industry originated in the province of Chou-Tang and the secret was guarded by Chinese for about 3000 years. First century BC is the period sericulture growth in Southern Europe. From China it spread to Korea during 1200 BC. In third century BC, Semiramus a Japanese General in the army of Empress Singu—Kongo invaded and conquered Korea and the prisioners of war brought the technology to Japan. First century BC is the period of sericulture growth in Southern Europe. 19th Century—Japan gave serious attention to the development of industry. Tibet King of Koten married Chinese Princess who brought silkworm eggs in her head dressings. From Tibet, Silkworm rearing spread to India & Persia during 140 BC.

Mulberry belongs to the genus Morus having more than 20 species. *Morus alba* is native of Indo-Himalayan region and a perennial tree. In peninsular India, it is maintained mainly as bush and to certain extent as tree. The leaves are simple alternate stipulate and serrated. Mulberry is unique in possessing both lobed and unlobed leaves on the same twig (heterophylly).

The inflorescence of mulberry is a catkin bearing unisexual flowers. The female floret has a pistil and four perianths / sepals. The pistil is composed of stigma, style and ovary. The male floret has four stamens and four perianths. The stamen is composed of anther and filaments. The plant is generally dioecious and rarely monoecious.

Mulberry can be grown upto 800 m MSL. For the optimum growth of mulberry and good sprouting of the buds, the mean atmospheric temperature should be in the range of 13°C to 37.7°C. The ideal temperature should be between 24 to 28°C with relative humidity of 65 to 80 percent and sunshine duration of 5 to 12 hours per day.

Mulberry can be grown in a rainfall of 600mm to 2500mm. Under low rainfall conditions, the growth is limited and requires supplemental irrigation. On an average, 50mm once in 10 days is considered ideal for mulberry. Mulberry Genotypes (Morus spp. ; Family: Moraceae) The important characteristic feature of the members of the family Moraceae is the presence of idioblast which is nothing but the enlarged epidermal cell in the leaf. Some important species of mulberry are as follows:

White mulberry	-	*Morus alba*
Black mulberry	-	*M. nigra*
Red mulberry	-	*M. rubra*
Russian mulberry	-	*M. tartarica*
Indian mulberry	-	*M. indica*
Creamy mulberry	-	*M. serata*

The soil should be deep fertile, well drained clayey loam in texture, friable, porous and with good moisture holding capacity. Slightly acidic soils (6.2 to 6.8 pH) free from injurious salts are ideal for good growth of mulberry plant, Saline and alkaline soils are not preferred.

Mulberry Plants Propagation

The method of propagation of mulberry plants is classified into two types viz., sexual and asexual. The sexual method of propagation is used mostly for breeding purpose and for the production of triploids. Since mulberry is mostly cross pollinated crop, true to type of plant materials cannot be assured. In the asexual method of propagation, there are three methods viz., grafting, layering and cuttings.

Grafting and layering methods are not popular because of the labourious process, high cost and skill involved in it. Moreover large number of planting materials cannot be obtained in short period of time. In cuttings planting, it may be planted straight away in the field itself or nursery may be raised and the sprouted and rooted saplings may be planted in the mainfield. The latter method is advisable because of its easy establishment in the field.

Methods of Planting

Material Selection

Generally the mulberry plants are raised from semihard wood cuttings. The cuttings are selected from well established plants of 8-12 months old. Only full grown thick main stems, free from insect and disease damages having a diameter of 7-10 mm are chosen for preparation of cuttings. The cutting should be of 15-20 cm with 3-4 active buds and should have 45° slanting cut at the bottom end. Care should be taken to make a sharp clean cut at both the ends of cuttings without splitting the bark. Manually/power operated mulberry cutters (stem cutting machine) is available for quick cutting of propagative material.

Nursery Bed Preparation

Select 800 sq.m. of red loamy soil near water source for raising saplings for planting one hectare of main field. Apply 1600 kg of FYM @ 20 t/ha) in the nursery area. Mix well with the soil. Raise nursery beds of 3m x 1.75m size. The length may be of convenient size depending upon the slope, irrigation source, etc. Provide a drainage channel and avoid shady area.

Pre-treatment of Cuttings

Prepare a slurry of Azospirillum. Mix one kilogram of Azospirillum culture in 40 litres of water. Keep the bottom end of the cuttings for 30 minutes in the slurry before planting.

Nursery Planting

Apply VAM at 100 g / m^2 of nursery area. Irrigate the nursery bed. Plant the cuttings in the nursery at 15 cm x 7 cm spacing at an angle of 45°. Ensure exposure of one active bud in each cutting.

Nursery Management

- Irrigate the nursery once in three days.
- Dust one kg of any one of the following chemicals around the nursery bed to avoid termite attack.
    - Endosulfan
    - Malathion
    - Quinalphos

To avoid root rot and collar rot, drench the soil with carbendazim (2g/1) using rose can.

Age of Sapling

The saplings are ready for transplanting in the mainfield after 90-120 days of planting.

Land Preparation

— Bring the soil to fine tilth by repeated ploughings and harrowing

— Plough with mould board plough followed by country plough

— If hardpan is present, use chisel plough at 0.5 m distance in perpendicular directions.

Organic Manure

— Apply organic manure (FYM) @ 20 t/ha for the irrigated crop and 10 t/ha for rainfed crop

— Apply the organic manure (FYM) during last plough for incorporation

— Apply organic manure (FYM) or compost 1.25 kg/ pit in case of pit system of planting

Planting Method

— *Row System*: Form ridges and furrows at 60 cm for irrigated crop and 90 cm for rainfed crop

— *Pit System*: After field preparation, make pits of 45 x 45 x 45 cm size at 90 x 90 cm spacing

— *Kolar System*: It is similar to the row system except the distance between the plants is very much less. Here the distance between rows is 30 to 45 cm. The Kolar system is normally practiced in different districts of Karnataka.

— *Strip System*: This is popular in West Bengal. In this system the distance between plants is 10 to 15 cm.

Here close system of mulberry cultivation is followed. However, mulberry is grown in strips. If each strips consists of two rows, it is called as Dothaki. If each strips consists of three rows then it is called as Thethaki.

— *Angular System*: This is normally practiced in slopes of Nilgris. In this system the distance between plant is similar to pit system. Here the distance between the strips in 90 cm. In this method intercultivation is easier. Plants in adjacent rows are planted in such way that they form a triangular arrangement.

— *Paired Row System*: Plant the cuttings / saplings at a spacing of 120 cm between paired rows and 75 cm with in paired rows. Intercrops can be raised in the wider space.

Planting Time

— Plant during rainy season

— Avoid planting during winter and summer months

Fertiliser Management

Irrigated / Semi-irrigated

	Row system			Pit system		
	N	P (kg/ha)	K	N	P (kg/ha)	K
Recommendation	300	120	120	280	120	120
Split doses						
First crop	60	60	60	60	60	60
Second crop	60	--	--	40	--	--
Third crop	60	60	60	40	--	--
Fourth crop	60	--	--	60	60	60
Fifth crop	60	--	--	40	--	--
Sixth crop	--	--	--	40	--	--

— Apply fertilisers as per soil recommendation

— Apply the first dose of fertilisers three months after planting

— Follow subsequent fertiliser application after each leaf harvest and pruning

— Apply straight fertilisers to minimise the cost

Rainfed

	N	P (kg/ha)	K
Recommendation	100	50	50
First dose	50	50	50
Second dose	50	--	--

— Apply the first and second doses coinciding with South West and North east monsoons respectively.

Biofertilisers

— Apply Azospirillum @ 4 kg/ha along with phosphobacterium @ 15 kg/ha

— Mix the biofertiliser with 50 kg of FYM for uniform distribution

— Apply the biofertilisers once in six months

— Ensure irrigation after application

— Do not mix biofertilisers with inorganic fertilisers

Deficiency Symptoms

Identify the deficiency symptoms in the field with the following clues.

Major nutrients

Nitrogen: Young leaves show chlorosis with necrotic

spots, poor branching, lack of vigour, slender stem and stunted growth.

Phosphorus: Intravenal chlorosis of older leaves, chlorosis spreads throughout the leaves followed by marginal necrosis and defoliation, slender stem with stunted growth of roots, without fresh growth.

Potassium: Young leaves show marginal scorching. Coarse and non-juicy leaves with necrosis.

Secondary nutrients

Calcium: Defoliation of younger leaves with necrosis along the veins, stem becomes woody and short with yellowish tips, stubby and dry roots

Magnesium: Chlorotic leaves with white tip.

Sulphur: Stunted plant growth with mild chlorosis of leaves, yellowing of young leaves

Micro nutrients

Zinc: Stunted plant growth, chlorosis of young leaves, complete bleaching of midrib, bottom leaves with bleached spots.

Copper: Leaf tip becomes white, narrow and twisted, stunted growth, dieback of terminal bud, marginal chlorosis.

Iron: Typical intervenal chlorosis, papery white leaves with necrotic spot.

Boron: Rosetting of terminal growth, death of terminal bud and cracked stem.

Molybdenum: Pale yellow leaves and poor growth, undulating leaf surface, necrotic spots along midrib.

Manganese: Thin foliage, dark green bands along the midrib; light green stripe between dark green veins, twigs with dieback symptom.

Apply recommended major/secondary nutrients based on the deficiency symptoms. For micro nutrients according to the deficiency symptom expressed, apply micronutrients as foliar spary @ zinc sulphate 5 g, Ferrous sulphate 10 g, Borax 2.5 g, copper sulphate 2.5 g, manganese 2.5 g or sodium molybdate 100 mg/lit of water on 20th and 40th day after pruning using high volume sprayer (Spray fluid 500 lit/ha). Add wetting agent, teepol @ 0.5 ml/lit. for better adherence on the foliage.

Water Management

The mulberry crop requires about 125 to 150 mm of water over a period of 12 months. The water use efficiency in mulberry is about 20 kg. It will vary according to varieties and method of cultivation. Irrigation is done once in 15 days for clay soil, once in 10 days for clayey loam, and once in 7 days for sandy soil.

Irrigation Methods

Ridges and furrows method

— Most efficient method of irrigation

— Comparatively requires less amount of water

— The furrows serve as drainage channel during heavy rainfall.

Flat bed method

— Rectangular beds and channels are formed

— Water run off is relatively low
— More land is wasted in bunds and requires more labour for field preparation.

Sprinkler and drip irrigation

— Most efficient in water use
— Can be used in undulating terrains
— Fertilisers can also be applied along with irrigation water
— Initial cost is very high

Weed Control

The common weed flora in the mulberry garden is given below

Monocotyledonous Weeds (Grassy weeds)

— *Cyperus rotundus* (Korai / Nut grass)
— *Cyanodon dactylon* (Arugampull / Bermuda grass)

Dicotyledonous Weeds (Broad leaves)

— *Abutilon indicum* (Thuthi)
— *Trianthema portulacastrum* (Saranai)
— *Tridox procumbens* (Manjapoo)
— *Boerhaevia diffusa* (Mookaratai)
— *Amaranthus viridis* (Kuppaikerai)
— *Croton sparsiflorus* (Naimilakai)
— *Acalypha indica* (Kuppaimeni)
— *Parthenium hysterophorus* (Vizachedi)
— The weeds compete for water, nutrient and space

— They act as alternate host for many pests and diseases of mulberry plant.

Control Measures

Cultural

— Remove the stubbles and roots of weeds while preparing the land

— Use well decomposed manure to avoid dissemination of weeds

— Clean the implements before use

— Keep the channels free of weeds

Mechanical

— Operate country plough after pruning in the inter space

— Remove the weeds by hand hoe

Chemical Methods

— Use simazine @ 300 g or Diuron @ (2 lit/ha) in 2000 lit. water/ha as preemergence application.

— As post-emergence application use Paraquat (Grammoxone) @ 2-3 lit/ha.

— Spray Glycel @ 10 ml + 2, 4-D at 5 mg/lit of spray fluid as post emergence application. A total of 600 litres of spray fluid is required/ha.

— Use flood jet / fan type nozzle for spraying weedicide. Apply the weedicide immediately after pruning or within 2-3 days after pruning

Intercropping

Intercropping with short duration pulse crop enriches the

soil, gives additional revenue and also controls the weed growth. Anyone of the following crops can be grown as intercrop.

Black gram -	Co 5, VBN 1
Green gram-	Co 5, Paiyur-1
Cow pea -	Co 4, Co 5, Pusa-152

Mulching

Mulching with pruned mulberry twigs and other materials like straw and dried leaves will have the following advantages

— Controls weed growth

— Conserves soil moisture by reducing run-off

— Increases the infiltration of water

— Reduces the soil temperature

Pruning

Pruning is the methodical removal of branches of mulberry plant with the object of giving a convenient shape and size to increase the leaf yield and to improve its feeding value. Pruning of mulberry plant is also useful in adjusting the production period to synchronise with the leaf requirement for silkworm rearing and also to extend the leaf production period throughout the year.

Bottom Pruning: In this method, the plants are cut at ground level leaving 10-15 cm of stump above the ground. This type of pruning is done once in a year.

Middle Pruning: In this, the branches are cut at 40-60 cm above the ground level. After bottom prunings, subsequent cuts are made at 45-50 cm height.

Kolar or Strip System: In closely planted area, this type of pruning is done. In this system, the branches are cut at ground level every time. Thus, it receives five prunings every year. This type of severe pruning requires heavy fertilisation and irrigation.

Leaves Harvesting

The method of leaves harvest depends on the type of rearing practiced. There are three methods of harvesting mulberry leaves viz. i) Leaf picking ii) Branch cutting and iii) Whole shoot harvest.

Leaf Picking: In this method, individual leaves are harvested with or without petiole. Leaf picking starts 10 weeks after bottom pruning and subsequent pickings are done at an interval of 7 - 8 weeks.

Branch Cutting: In this method, the entire branches are cut and fed to the worms. Before that, topping is done to ensure uniform maturity of the lower leaves.

Whole Shoot Harvest: In this method, the branches are cut at ground level by bottom pruning. Shoots are harvested at an interval of 10-12 weeks and thus 5 to 6 harvests are made in a year.

Advantages

— Labour saving method

— Leaves are utilised to the maximum

— It does not require rearing equipments like trays and shelves

— The moisture is retained for a longer period and quality of the leaves is better

It is preferable to harvest the leaves during morning hours. Use wet gunny bags to store the leaves or cover the bamboo basket with wet gunny bags to keep it cool and fresh.

Insect Pests

Sucking Pests

1.Pinkmealy bug	*Maconellicoccus hirsutus (Green) Pseudocoocidae, Hemiptera*
Type of damage	Leaf yield and plant height are drastically reduced, Nymphs and adults cause damage
Symptoms	Apical shoots retarded, flattening of apical shoots, wrinkling of the affected leaves, become dark in colour. Symptoms are collectively called as Tukra (Bushy top) disease
Season	Prevalent throughout the year with higher incidence during summer

Management

— Cutting the affected shoots and burning

— Spray Fish Oil Rosin Soap (FORS) @ 40 g/l or dichlorvos 76 WSC @ 2 ml/l, (safer waiting period: 15 days)

— Release Cryptolaemus montrouzieri @ 300 beetles/acre and Scymnus coccivora @ 400 beetles/acre.

2. Jassid/Leafhopper	*Empoas caflavescens(F.) Cicadellidae, Hemiptera*
Type of damage	Adults and nymphs attack the mulberry leaves from lower side of the margin of the veins and deplete leaf nutritive value

Symptoms	Hopper burn Triangular dark brown spot at the tip and margin of leaves: Drying starts from periphery to midrib of leaf. Finally leaves become cup shaped and wither
Season	Summer

Management

Spray dichlorvos 76 WSC @ 2ml/l

3. *Black Scale:*	*Saissetia nigra (Nietm):Coccidae, Hemiptera*
Type of damage	Desapping the leaves, shoots dry up from the tip
Symptom	Affected shoot studded with black scales, yellowish or mottled appearance of leaf blade.
Season	Summer months

Management

Swabbing the stem with diesel oil and soap emulsion (1:3 ratio) after pruning. Scrapping with a plate to dislodge the insects. Spray malathion @ 2 ml/l.

4. *Thrips*	*Pseudodendrothrips mori Niwa Thripidae, Thysanoptera*
Type of damage	Injure the epidermal tissues and desap. Early maturity, depletion of moisture, reduction in crude protein and total sugars are met. Leaves are unfit to rear silkworm
Symptoms	Leaves show white streaks, finally yellowish blotches on maturity

Season	Throughout the year, more during summer

Management

Spray dichlorvos 76 WSC @ 2ml/lit.

5. *Spiralling white fly*	*Aleurodicus ' isperses Russell Aleyrodidae, Hemiptera*
Type of damage	Desapping the leaves Depletes nutritive value of leaves Nymphs and adults cause damage
Symptoms:	
	— Adults and nymphs congregate on the lower surface of leaves — Desap the leaves, results in yellowish speckling on leaves — Leaves crinkle and curl and sooty mould appears — Infestation spreads from the bottom leaves to the top — Eggs are laid on lower leaves with irregularly spiralling deposits of waxy white flocculence
Season	Throughout theyear, severe during winter
Important alternate hosts	Guava, bhendi, tapioca, papaya

Management

- — Collection and destruction of leaves with egg masses, nymphs and adults
- — Setting up yellow sticky traps
- — Removal of weeds
- — Spray Fish Oil Rosin Soap @40 g/l or dichlorvos 76 WSC @ 2 ml/lit. e) Taking management practices including the alternate hosts (crops) simultaneously.

Foliage Feeders

1. Tobacco Leaf Caterpillar, Cutworm Spodoptera litura (F) Noctuidae; Lepidoptera

Type of damage	Larvae cut the shoot of young plants. The cut portion of the shoot dries up and falls down. They also feed upon the leaves. Larval stage is destructive.
Symptoms	Newly sprouted mulberry or young plants are found with cut branches with dried leaves. Chlorophyll scrapping and defoliation symptoms in older leaves
Season	August to February

Management

- — Deep ploughing to expose the pupae
- — Setting up light traps and pheromone traps
- — Collection and destruction of leaves with egg masses and larvae
- — Spray dichlorvos 76 WSC @ 2 ml/l

— Growing trap crop as on the bunds

— Avoid spraying of Bt formulations on mulberry to manage lepidopteran pests since which may harm the silkworms.

2. *Hairy Caterpillars*

a) *Eupterote mollifera* W. (Eupterotidae) (Moringa hairy caterpillar)

b) *Euproctis fraterna* (Moore) (Lymantridae) (Tussock caterpillar)

c) *Porthesia scintillans* (Lymantridae) Brown hairy caterpillar/Tussock moth

Type of damage	Reduction in leaf yield. Larvae cause the damage. E. mollifera larvae rest in group in branches during day and defoliate the leaves during night hours.
Symptom	Branches of mulberry plants without leaves
Season	February to August

Management

a) Install light traps.

b) Collection and destruction of egg masses and gregarious young larvae.

c) Deep ploughing and flood irrigation.

d) Spray dichlorvos 76 WSC @ 2 ml/Avoid plucking of leaves for 15 days after spraying

e) Fish Oil Rosin Soap (40 g/l) on trunk during day to kill the resting *E. mollifera*-larvae.

3. *Leaf webber: Glyphodes sp. Pyraustidae, Lepidoptera*

Type of damage	Larvae defoliate the mulberry plants. Reduction in leaf yield
Symptomsn	Larvae web the leaves together feeding inside, skeletonise them. Cause damage by folding the leaves and by webbing the tender shoots. Apical tips are preferred for feeding.
Season	October - February

Management

— Clipping off affected terminal shoots

— Setting up light traps

— Spray dichlorvos 76 WSC @ 2 ml/l

4. *Ash weevil*

Myllocerus spp. Curculionidae, Coleoptera

M. subfusciatus G.,

M. viridanus

M. discolor B.

Type of damage:

— Adults and grubs cause injury to mulberry

— Adults feed on the leaves and buds Grubs feed on the rootlets

Symptoms

— Irregular notching on leaf margins

— Under severe attack, the plants wilt and dry up

Season July to October

5. *Grasshoppers*

- Neorthacris acuticeps nilgriensis Uvarov Acrididae, Orthoptera
- Cyrtacanthacris ranacea Acrididae, Orthoptera

Type of damage	Reduction in leaf yield—nymphs and adults cause damage
Symptoms	Branches of plants without leaves
Season	July—October

Management

- Deep ploughing and trimming the bunds during summer b) Spray with malathion 50 EC @ 4 ml/l or dichlorvos 76 WSC @ 2 ml/l.

Internal Feeders

1. *Stem girdler beetle* — *Sthenias grisator L. Cerambycidae; Coleoptera*

Type of damage	Reduction in plant growth and leaf yield. Adults and grubs cause the damage
Symptoms	The beetle cut around the main stem or branch leaving a clear girdle. The portion above' the girdle gradually wilts and dies. The grub tunnels into the wilting branches and feeds
Season	Throughout the year

Management

- Cutting and burning of the affected branches and stem

— Swabbing at the base of main stem or branches with 0.1 per cent malathion 50 EC.

2. *Termite / White ant* *Odontotermes sp. Termitidae, Isoptera*

Type of damage: Workers of the colony cause the damage. Drying of plants and reduction in leaf yield. Mulching with dry twigs favours the population build up in endemic area.

Symptoms: Feed upon roots and bark of young and old plants. Damaged plants dry up and die.

Season: October—February

Management

— Location and destruction of termite colonies by removing queen termite

— Treatment of mounds with 50 ml chlorpyriphos 20 EC

— Swabbing or drenching of established plants at the base with 50 ml chlorpyriphos 20EC.

Nematode Pests

1. *Root knot nematode:* *Meloidogyne incognita (Kofoid and White)*

Order: Tylenchida,

Class: Secernentia,

Phylum: Nematoda

Symptoms: Stunted plants, marginal necrosis and yellowing of leaves, formation of characteristic knots or galls on the roots.

Management

— Deep ploughing in summer

— Apply Neem cake @ 250 kg/ha

— Apply carbofuran 3G @ 30 kg/ha/year in four split doses (Safer period is 50 days)

Diseases

Fungal Diseases

1. *Leaf spot*	*Cercospora moricola (Cooke)*
Family	Dematiaceae,
Order	Moniliales, Class: Hyphomycetes,
Sub division	Deuteromycotena
Symptoms	Brownish irregular leaf spot in the initial stage, holes are seen in later stage Severely affected leaves become yellowish and fall off prematurely
Season	June to December

Management

Spray carbendazim @ 500-625 g/ha

2. *Powdery mildew*	*Phyllactinia corylea (Pers)*
Family	Erysiphaceae,
Order	Erysiphales,
Class	Plectomycetes
Subdivision	Ascomycotena

Symptoms	Initially white powdery patches on lower surface of leaves and later turn black to brown in colour. High humidity and low temperature favour the outbreak of the disease.

Management

— Provide wider spacing

— Grow resistant varieties like MR 2 and China white

— Spray carbendazim @ 500-625 g/ha

— Releasing fungivorous lady bird beetles.

3. *Leaf Rust*	*Cerotelium Fici (Cast)*
Family	Pucciniaceae,
Order	Uredinales,
Class	Teliomycetes
Subdivision	Basidiomycotena Symptoms Presence of brownish to black eruptive lesions/ spots on the leaf surface. Leaves become yellowish and wither off prematurely.

Management

— Provide wider spacing

— Spray carbendazim @ 500-625 g/ha

4. *Root Rot:*

Black root rot	*Macroph omina phaseolina*
Family	Sphaeropsitaceae,
Order	Sphaeropsitales,
Class	Coleomycetes

Subdivision	euteromycotena
White root rot	*Rosellinia necatrix Berlese*
Order	Sphaeriales
Symptoms	Plants become weak, leaf buds grow feebly, leaves wither off and plants die. The stump region seen with whitish rey mycelial mat.
Violet root rot	*Helicobasidium mompa Tanaka*
Order	Tremellales,
Class	Basidiomycetes
Symptoms	Withering of leaf and dying of the plant. Epidermal tissues of the roots with violet coloured mycelial mat.

Management

a) Uproot the infected plant and the stump and root portions are burnt.

b) Application of Neem cake @ 1 tonne/ha in 4 split doses.

c) Application of antagonistic fungus Trichoderma viride @ 25 g / plant

d) Drenching the soil with carbendazim @ 10 ml of 1 % concentration per plant.

5.*Stem canker*	*Lasiodiplodia (Botryodiplodia) theobromae*
Family	Sphaeropsitaceae,
Order	Sphaeropsidales,
Class	Deuteromycetes
Subdivision	Deuteromycotena

Symptoms	*Nursery*: Failure of cuttings to sprout. Sudden withering and death of sprouts. Discolouration and drying of stems and buds above the soil. Rotting and peeling of bark on stem below the soil surface. Black mycelial threads seen below infected bark and black eruptions on the bark of the infected stem portion. *Grow-up plants*: Greyish brown discolouration of the bark at the cut ends of the stem. Delayed sprouting, death of buds and sprouts, black eruptions on the bark in the infected region and death of plants. The above symptoms can be observed few days after the plants are pruned.
Season	Throughout the year, more severe during winter months.

Management

— Exclusion of the inoculum

— Roguing

— Field sanitation

— Land use

— Planting season—avoid planting in winter months

— Proper pruning

— Chemical control

i) Pre-treatment of cuttings with carbendazim @ 4 g/l for a period of 12 hours.

ii) After pruning, the cut surfaces of the stems should be dressed with a spray/smear of carbendazim @ 4g/1.

Bacterial Disease

Bacterial blight: Psesudomonas mori Boyar and Lambert

Family	Pseudomonadaceae,
Order	Pseudomonadales,
Class	Schizomycetes
Symptoms	Numerous irregular water soaked patches on the lower surface of leaf. Leaves become curled, rotten and turn brownish black in colour. Black longitudinal lesions are seen on the bark of young shoots.
Season	Rainy

Management

— Uprooting and burning

— Spray 0.1 per cent of Streptomycin or Streptocyclin (Safer period is 15 days)

7

Mulberry Cultivation in India

In India, most states have taken up sericulture as an important agro-industry with excellent results. The total acreage of mulberry in India is around 282,244 ha.

Table 1. Area under mulberry cultivation in different states

State	Area (ha)
ANDHRA PRADESH	38,084
ASSAM	2,813
JAMMU & KASHMIR	4,717
KARNATAKA	166,000
KERALA	1,164
MADHYA PRADESH	2,043
MANIPUR	*25,975
TAMIL NADU	9,491
UTTAR PRADESH	5,665
WEST BENGAL	21,358
OTHER	4,934
TOTAL	282,244

Species and Varieties

There are about 68 species of the genus *Morus*, the majority of them occur in Asia, especially in China (24 species) and Japan (19). Continental America is also rich in its *Morus* species. The genus is poorly represented in Africa, Europe and Middle East, and it is not present in Australia.

In India, there are many species of *Morus*, of which *Morus alba, M. indica, M. serrata* and *M. laevigata* grow wild in the Himalayas. Several varieties have been introduced belonging to *M. multicaulis, M. nigra, M. sinensis* and *M. phillippinensis*. Most of the Indian varieties of mulberry belong to *M. indica*.

Table 2. Mulberry varieties under cultivation in different states

Variety	Region	Developed at	Origin
Kanva-2	South India Irrigated	CSRTI, Mysore	Selection from natural variability.
S-36	"	"	Developed through EMS treatment of Berhampore Local.
S-54	"	"	"
Victory-1	"	"	Hybrid from S30 x Ber C776.
DD	"	KSSRDI, Thalaghattapura	Clonal selection.
S-13	South India Rainfed	CSRTI, Mysore	Selection from polycross (mixed pollen) progeny.
S-34	"	"	"
MR-2	"	"	Selection from open pollinated Hybrids.

Contd....

Contd....

S-1	Eastern and N.E. India Irrigated	CSRTI, Berhampore	Introduction from (Mandalaya) Burma.
S799	"	"	Selection from open pollinated Hybrids.
S1635	"	"	Triploid selection.
S146	N.India and Hills of J & K Irrigated	"	Selection from open pollinated hybrids.
Tr-10	Hills of Eastern India	"	Triploid of Ber.S1.
BC 259	"	"	Back crossing of hybrid of Matigare local x Kosen with Kosen twice.
Goshoerami	Temperate	CSRTI, Pampore	Introduction from Japan.
Chak Majra	Sub-temperate	RSRS, Jammu	Selection from natural variability.
China White	Temperate	CSRTI, Pampore	Clonal selection.

In China there are 15 species. Out of which, 4 species, *Morus alba, M. multicaulis, M. atropurpurea* and *M. mizuho* are cultivated for sericulture. In the ex-USSR *Morus multicaulis, M. alba, M. tartarica* and *M. nigra* are present.

Though mulberry cultivation is practiced in various climates, the major area is in tropical zone covering Karnataka, Andhra Pradesh and Tamil Nadu states, with about 90%. In the sub-tropical zone, West Bengal,

Himachal Pradesh and north-eastern states have major areas under mulberry cultivation. The details of mulberry varieties under cultivation in different states of India is given in Table-2.

Mulberry is a fast growing deciduous woody perennial plant. It has a deep-root system. The leaves are simple, alternate, stipulate, petiolate, entire or lobed. Number of lobes varies from 1 to 5. Plants are generally dioecious. Inflorescence is catkin with pendent or drooping peduncle bearing unisexual flowers. Inflorescence is always auxiliary. Male catkins are usually longer than the female catkins.

Male flowers are loosely arranged and after shedding the pollen, the inflorescence dries and falls off. Number of parianth lobes are 4. Number of stamens are 4 and implexed in bud. Female inflorescence is usually short and the flowers are very compactly arranged. Number of parianth lobes are 4 and persistent. Ovary is one-celled and stigma is bifid. The chief pollinating agent in mulberry is wind. Fruit is a sorosis and the colour of the fruit is mainly violet black.

Most of the species of the genus Morus and cultivated varieties are diploid having 28 chromosomes. However, triploids (2n=(3x)=42) are also extensively cultivated for their adaptability, vigorous growth and quality of leaves.

Climatic Requirement

Mulberry thrives under various climatic conditions ranging from temperate to tropic located north of equator between 28° N to 55°N latitude. The ideal range of temperature is from 24-28°C. It grows well in places with annual rainfall ranging from 600mm to 2500mm. In area with low rainfall, the growth is limited due to moisture stress resulting in low yields. On an average mulberry requires 85,000 gallons of water per hectare once every 10 days in case of loamy soils and 15 days in clayey soils.

The atmospheric humidity in the range of 65-80% is ideally suited for mulberry growth. Sunshine is one of the important factors controlling growth and leaf quality. In tropics, mulberry grows with a sunshine range of 9.0 to 13.00 hours a day. Mulberry can be cultivated from sea level up to an elevation of 1000 m above mean sea level.

Soil Condition

Mulberry tlourishes well in soils which are flat, deep, fertile, well drained, loamy to clayey, porous with good moisture holding capacity. The ideal range of soil pH is 6.2 to 6.8. The optimum pH required for mulberry is 6.5 to 6.8. Soil amendments may be used to correct the soil to get required pH. The quantity of gypsum or lime to be applied in different cases are as mentioned below:

Soil amendments: (To be taken up after soil test)

PH range (To bring pH to 6.8)	Quantity of gypsum/ha
7.4 to 7.8	2.0 MT
7.9 to 8.4	5.0 MT
8.5 to 9.0	9.0 MT
9.1 to above	14.0 MT

pH range quantity of lime/ha (to bring pH to 6.8)

pH range	Quantity of lime/ha (To bring pH to 6.8)		
	Plain	Hilly areas	Soil type
5.5 to 6.5	1.25 MT	2.5 MT	Sandy
	2.50 MT	5.0 MT	Sandy loamy
	5.0 MT	7.5 MT	Loamy
	7.5 MT	8.75 MT	Clay loamy

Method of application: The powdered gypsum/lime is mixed well with soil with soil of the garden and irrigated to stagnation for 48-72h. Later the water is leached out by drainage and dried (suitable for ploughing and intercultural operations).

Mulberry Cultivation in South India—Karnataka, Andhra Pradesh and Tamil Nadu

Suitable Mulberry Varieties

Kanva-2, S-13 and S-34 varieties are recommended for rainfed (rainfall : 500-800 mm) regions of South India.

Kanva-2

Belongs to *Morus indica* - Diploid. Widely cultivated in Southern India. Selection from natural population of Mysore Local variety.

Flower and sorosis: Female, profuse flowering, many sorosis.

Economic characters: Medium leaf maturity, yields about 30 to 35 MT/ha/year under irrigated condition. Leaf moisture content 70%, protein content 21% and sugar content 11.5%. High rooting ability (80%) and wide adaptability.

Resistant to leaf spot. Moderately resistant to leaf rust and powdery mildew.

S-13

Belongs to M. indica. Selection from open pollinated hybrids of Kanva-2. Recommended for rainfed areas of South India during 1990.

Flower: Male, profuse flowering.

Economic characters: Yields 8-12 MT/ha/yr under rainfed conditions depending on rainfall. Moisture content 70.6% protein content 24.3% and sugar content 13.8%.

Resistant to leaf spot and powdery mildew, moderately resistant to leaf rust and tukra infestation.

S-34

Belongs to *M. india* - Diploid. Selection from progeny of S30 x Berc 776. Recommended during 1990 for rainfed areas with black cotton soils of South India.

Flower and sorosis: Male, profuse flowering, occasionally few sorosis.

Economic characters: Under rainfed conditions, yields about 15 MT/ha/yr. Moisture content 70%, crude protein content 23.7%. Soluble sugar content 13.2%.

Resistant to powdery mildew and leaf rust. Moderately resistant to leaf spot and susceptible to tukra infestation.

Establishment of mulberry

i) *Preparation of land*: The land meant for mulberry cultivation is ploughed deep with heavy mould board plough up to a depth of 30-35 cm. Thereafter the land is repeatedly ploughed 2-3 times with a country plough to bring the soil to a fine tilth. The land should be properly levelled if it is sloppy. A basal dose of well decomposed Farm Yard Manure or compost is applied at the rate of 10 MT/ha and thoroughly incorporated into the soil.

ii) *Spacing*: The spacing commonly followed for a rainfed garden is 90 x 90cm Pits of 35 x 35cm. are prepared. About 1 kg farm yard manure/pit should be added.

iii) *Preparation of cutting and planting*: Branches of 8-10 months old and about 50 mm. in diameter should be used for preparation of cuttings of 22-25 cm. length with 5-6 healthy buds. Three cuttings are planted/pit in a triangular form with a spacing of 15 cm. leaving only one bud exposed above soil surface. If planting is done with saplings, then one sapling is sufficient/pit. Planting should be done during June /July after onset of monsoon.

iv) *Inter cultivation*: During first year of plantation, inter-cultivation should be done manually. Once the mulberry plants are established, bullock ploughing is carried out.

Application of fertilizer: 50N:25P:25K Kg/ha/yr in two doses.

First dose Suphala (15:15:15) 167kg, after 2 months of planting Second dose Urea 55 kg orCAn (100kg) or Ammonium sulfate (125kg), end of September or early October before cessation of monsoon rains

v) *Pruning and leaf harvest*: First crop should be taken 6 months after plantation when mulberry gets well established. Then onwards, two more crops are taken during first year by leaf picking method. Mulberry should be pruned after the completion of one year at the onset of next monsoon. The pruning is carried out with a sharp sickle or a pruning saw at the height of 25-30 cm. from the ground.

vi) *Green manuring and cover mulching*: Green manure crops can be grown as inter-crop with mulberry. It should be done during monsoon only. The green manure crops (cow pea, horse gram, dhaincha) should be incorporated into soil by ploughing before the flowering starts and well before rains cease. After that the plots may be given cover mulching with any dry material or weeds in which seed is not a source of multiplication.

Mulberry Under Rainfed Condition

Inputs required for rainfed garden (per ha. per year) Recommended inputs for gardens maintained under rainfed condition.

Spacing : 90 cm. x 90 cm.

1. FYM/compost, 10 MT, single dose at the onset of monsoon
2. Azotobacter biofertilizer, 4 kg/crop, two times/yr (during rainy season)
3. VAM inoculum, 1000 kg, once in life span of mulberry (inoculation through maize rootlets)
4. Suphala, 167 kg, I crop
5. Single Super Phosphate, 156 kg, I crop
6. Muriate of potash, 42 kg, I crop
7. Urea (55 kg) or Cam (100kg), III crop
8. Green manuring, 15 kg

Crops like horse gram, cow pea, sun hemp, dhaincha, etc., should be incorporated into soil by ploughing before flowering and cessation of monsoon.

Leaf Harvest: Individual leaf harvesting should be made. The yield for different varieties is as mentioned below:

K-2 – 10-12 MT/ha/yr
S-13 – 14-15 MT/ha/yr
S-34 – 14-15 MT/ha/yr

Mulberry Cultivation under Irrigated Condition

Suitable Mulberry Varieties

Kanva-2, S-36, S-54, DD, MR2 (especially in Tamil Nadu) and V-1 varieties are recommended for irrigated conditions.

Kanva-2

Belongs to *Morus indica* - Diploid. Widely cultivated in Southern India after the same was recommended for cultivation in 1969 by CSR & TI, Mysore. Selection from natural population of Mysore Local variety.

Flower and sorosis: Female, profuse flowering, many sorosis.

Economic characters: Medium leaf maturity, yields about 30 to 35 MT/ha/year under irrigated condition. Leaf moisture content 70%, protein content 21% and sugar content 11.5%.

Resistant to leaf spot. Moderately resistant to leaf rust and powdery mildew. High rooting ability (80%) and wide adaptability.

S-36

Belongs to *M. indica*. Developed at CSRTI, Mysore and recommended during 1984. Evolved from Berhampore local by chemical mutagenesis. Cultivated in Southern India. Moderate rooting ability.

Economic characters: Yields about 38 - 45 MT/ha/yr under assured irrigated conditions of South India. Moisture content 76 % protein content 22 % and carbohydrate content 28 %.

Tolerant to leaf spot and powdery mildew. Moderately susceptible to leaf rust and to tukra infestation.

Most suitable to young age silkworm rearing. Sapling plantation recommended due to moderate rooting ability.

S-54

Belongs to *M. india*. Developed at CSRTI, Mysore and recommended during 1984. Selected from Berhampore local by chemical mutagenesis (EMS). Recommended for assured irrigated conditions of South India.

Economic characters: Yields about 45 MT/ha/yr under assured irrigated conditions. Highly responsive to agronomical input. Moisture content of leaf 70.5% protein content 23.9% and sugar content 13.8%. Moderately resistant to powdery mildew and leaf rust and resistant to leaf spot. Looses moisture very quickly. Good rooting ability.

DD : Selected from natural population of Dehra Dun variety and recommended by Karnataka State Sericultural Research and Development Institute, Thalaghattapura. Recommended for Southern India.

Morphology: Erect, thin branches, coarse leaves, greenish grey colour bark. Lower branches spreading, leaves unlobed, big size, ovate shape. Yields 35-40 MT/ha/yr under assured irrigation condition.

MR – 2

Belongs to *M. sinensis* – Diploid. Selection from open pollinated hybrid population. Developed at CSR & TI, Mysore and recommended for propagation in Tamil Nadu. Mainly cultivated in Tamil Nadu under both irrigated condition in plains and rainfed conditions in hilly regions.

Economic characters: Yields about 30-35 MT/ha/yr under irrigated conditions of Tamil Nadu. Moisture content 68%, protein content 23.2%, sugar content 13.2% (Fig. 7).

Resistant to powdery mildew disease, suitable for hilly areas.

V-1

Belongs to *M. indica* L. Recently developed from a cross of S-30 and Berc.776 at CSR&TI, Mysore. Recommended during 1996 for assured irrigated conditions.

Flower: Male, profuse flowering, occasionally few sorosis.

Economic characters: Yields about 70 MT/ha/yr under assured irrigated conditions.

Very high sprouting. Moisture content 78.9% and 72.5% in young and matured leaves respectively, protein content 24.6% and total sugar content 16.98%.

Moderately resistant to leaf rust and tukra infestation and resistant to leaf spot. Quick sprouting ability and very high rooting ability (> 94%) high photosynthetic rate and higher water use efficiency are additional advantages. Moreover, leaves are suitable for both young and grown bivoltine silkworm rearing.

Plantation of Mulberry

i) *Selection of site*: Mulberry flourishes well in soils which are flat, deep, fertile, well drained, loamy to clayey, porous with good moisture holding capacity. The ideal range of soil pH is 6.2 to 6.8. Mulberry can be grown in saline, alkaline and acidic soils after suitably amending the soils.

ii) *Preparation of land*: The land meant for mulberry cultivation is ploughed deep with heavy mould board plough up to a depth of 30 – 35 cm. Thereafter the land is repeatedly ploughed 2-3 times with a country plough to bring the soil to a fine tilth. The land should be properly leveled if it is sloppy. A basal dose of well decomposed farm yard manure or compost is applied at the rate of 20 MT/ha and thoroughly incorporated into the soil.

iii) Plantation can be raised by using both cuttings or saplings. The varieties ideally suited for irrigated conditions are Kanva-2, S-36 and V-1. Branches of 6-9 months old and about 15 ml.. in diameter should be used for preparation of cuttings of 15-18 cm. length having 3-4 healthy buds for raising nursery or for planting directly in the field.

iv) *Spacing*: Plant spacing of 90 x 90 cm. is ideal for mulberry. While taking Plantation directly, two cuttings/pit required to be used. In case of using saplings, only one sapling/pit is sufficient. Presently a paired row plantation with the spacing of (90+150)cm x 60cm, is recommended.

v) *Inter-cultivation*: Two months after planting, weeding is done. A second weeding is done after another 2-3 months. Thereafter, inter-cultivation should be done after every shoot or leaf harvest.

vi) *Irrigation*: The plantation should be taken up during the onset of monsoon to take advantage of the rains. If the rain is not sufficient, the land should be irrigated at regular intervals of 8-14 days depending on type of soil. About one and a half to two acre inches of water is required/irrigation.

vii) *Application of fertilizer*: The total dose of fertilizer to be applied in first year is 100 N: 50 P: 50 K/kg/ha/ yr. This is applied in two doses. The first dose is applied when the plantation is about 2 months old at the rate of 50 N: 50 P: 50 K/kg/ha. The second dose is applied after taking the leaf harvest at the rate of 50 kg N alone/ha.

viii) *Pruning*: After 6 months of plantation, mulberry attains a height of about 1.50 to 1.75 m. and is ready for harvest. The first harvest is taken by bottom pruning. The second leaf harvest is taken after 12 weeks of first leaf harvest and the third harvest is taken 12 weeks after the second harvest by shoot harvest. From second year onwards, harvest is made at an interval of 70 days by shoot harvest method.

Maintenance of Mulberry under Irrigated Conditions

Spacing 90 cm. x 90 cm. or (90 + 150) cm x 60 cm

Recommended inputs:

i) FYM/compost	20 MT	In two equal split doses (10 MT FYM+ MT vermi compost)
ii) Azotobacter	20 kg	In five equal split doses
iii) N-triacontanol***	250 ml	In two equal split doses/crop
iv) VAM inoculum** (for existing garden)	1000 kg	One dose in the life span of mulberry (inoculation through maize as host plant)
v) Ammonium* sulphate or Urea or CAN	750 kg 325 kg 600 kg	In five equal split doses
vi) Single super Phosphate	375 kgs	In two equal split doses(I and III crop)

Contd....

Contd....

vii) Muriate of potash	200 kg	In two equal split doses. (I and III crop)

* 50% of chemical fertilizer can be cut down due to application of azotobacter and VAM.

** VAM inoculation is not required for the garden, planted with Mycorrhiza inoculated saplings.

*** 1st spray: In between 10-15 days after pruning/leaf plucking 2nd spray : 10 days after Ist spray.

Leaf Harvest: Leaf harvesting is done either through individual leaf harvest or shoot harvesting. Later one is more economical and useful for shoot method of silkworm rearing. The yield of varieties is mentioned below:

Kanva-2	-	32-35 MT/ha
S-36	-	38-45 MT/ha
V-1	-	60-70 MT/ha

Mulberry Cultivation in Hilly Areas

Suitable Mulberry Varieties

S-1, S-7999, S-1635, S-146, Tr-10 and BC-259 varieties are recommended for hilly regions of north and north-eastern India.

Establishment of Mulberry

i. *Preparation of land*: If the land is having gentle slope, it can be leveled with minor land shaping and providing suitable type of bunds across the slope. If the slope is more, contour bunding terrace planting or contour line planting can be adopted. In more sloppy areas, platform for individual plants on contour lines is more suitable as the same involves less soil cutting.

ii. *Spacing*: Spacing for tree planting depends on soil topography, extent of land available for cultivation and training method. In case of gentle slope, 3'X 3', 5'x 5' may be adopted. In case of more sloppy land 10' x 10' can be adopted.

Pits are to be prepared for plantation. In case of deep textured loose soils, 45 x 45cm. and in had shallow soils 60cm x 60cm x 60cm) pits are to be prepared. For each pit, 5kg (one iron pan) of FYM or compost must be applied.

iii. *Planting*: Saplings of 5 months age with 5-6 feet is found suitable to plant during the regular onset of monsoon. One sapling/ pit should be planted. The saplings should be supported with stick to ensure straight growth.

Recommendation with Institute Technologies for Hilly Areas (Tropical high land)

Spacing 90 cm. x 90 cm.

FYM/compost	20 MT	In single dose in the month of Jan/Feb.
Azotobacter	7.5 kg	In 3 equal split doses.
VAM inoculum	1000 kgs	In single dose in the life span of mulberry
CAN	200 kg	In 3 equal split doses.
SSP	156 kg	In 2 equal split doses.
MOP	84 kg	In 2 equal split doses.

CAN: Calcium Ammonium Nitrate
SSP: Single Super Phosphate
MOP: Muriate of Potash

iv. *After care of plantation*: After one month, all the buds except top 5-6 should be removed carefully without damaging the bark. Weeds around the plant should be removed and regular pot watering should be given. After 3 months of planting, second weeding should be given and 25g of suphala/plant should be applied in trench and should be covered with soil. 2nd dose of fertilizer (25g urea/plant) should be given before cessation of monsoon. Plants must be protected from grazing.

Since the recommendation is general, the inputs like fertilizers and amendments may be applied on the basis of soil test reports.

Mulberry Cultivation in Temperate and Sub-Temperate Regions

The following varieties are found suitable for temperate and sub-temperate regions.

Temperate - Goshoerami, China white
Sub-temperate - Chak majra, S-146

i.	Temperate (Kashmir)	- FYM 20 tons/ha/yr (irrigated) After annual pruning July-August
	All types of plantation (bush, dwarf tree) maintained under irrigated condition.	
ii.	Temperate (Kashmir) NPK @ 300:150:150 ha/yr in two equal splits	- 1st dose in April-May - 2nd dose in June-July

The leaf yield of different varieties are as mentioned below:

Goshoerami 15-20 MT/ha/yr
China white 15-20 MT/ha/yr
Chak majra 20-22 MT/ha/yr

Mixed Farming

Mulberry can be successfully grown as an inter-crop (medium mixed tree) between the rows of tea/coffee as shade plants. By this besides providing shade to the tea/

coffee plants, good quantity of quality leaves can be obtained for silkworm rearing and as feed for cattle and goats. Further, the pruned shoots are a good source of firewood.

Insecticide/ Pesticide/ Fungicide and Other Inputs (ha/y) (if required)

i) DDVP (Nuvan) (P) Soap (Washing Soap) Non-detergent water)	1.25 ltr. Should be sprayed if required to control tukra/leaf roller.(in 480 l 5 kg To mix with DDVP solution (Not required for controlling leaf roller).
ii) Ladybird beetle (CSRT)	625 adults, as a biological control for mealy bug.
iii) Bionematicide (N) (180 days shelf life)	80 kg, in 3 split doses at an interval of 4 months during intercultural operations to control root-knot disease.
iv) Neem oil cake (N)	2 ton, in 4 split doses at an interval of 3 months during inter-cultural operations to control root-knot disease.
v) Raksha (120 days shelf life)	1 kg, for 100 plants with 50 kg FYM for root-rot
vi) Bavistine (F) (0.2%) (Carbendazim: a.i.-50 WP)	1 kg, dissolved in 500 l water and sprayed to prevent leaf spot and powdery mildew dieases (safe period – 2 days).
vii) Kavach (F) (Chloro-tholonil: a.i.-WP)	1 kg, dissolved in 500 l water and sprayed to prevent leaf rust (safe period – 2 days).
viii) Dithane M-45(Man-cozeb: a.i. – 75 WP)	1 kg, dissolved in 500 l water and sprayed to prevent leaf blight due to fungi and bacteria (safe period – 2 days).
ix) Zinc sulhpate (Agro grade)	4.4 kg, in 440 l of water to increase leaf quality and quantity per crop.
x) Glyphosate 41% (Agro grade)	8.5 l, in two split doses (i.e. 2nd & 4th crop) to control weeds in mulberry garden).

Mulberry can also be grown as an inter-crop for cultivation in the wider space existing between sources of coconut plantation. A survey has shown that mulberry is inter-cropped with coconut in Channapatna, Ramanagaram, Kanakapura and Bangalore areas of Karnataka state.

Animal Feeding Practices

Mulberry is appropriately known as "Kalpa Vruksha" as all the parts of the plant has many uses. It is essential to sericulture as the foliage constitute the sole feed of mulberry silkworm *Bombyx mori*. Mulberry (*Morus* spp) is a fast growing tree which for convenience of sericulture practices is maintained as a bush. It produces very large amounts of renewable bio-mass in the form of branches, shoots, leaves and fruits. If mulberry is used for silkworm rearing it is possible to get 30-35 MT/ha of leaf every year. By growing mulberry, a farmer gets fodder, fuel and fertilizer.

With regard to fodder for animals, farmers in India feed their cows and goats with left over branches and leaves during silkworm rearing period. Many farmers feed their animals with full twig of mulberry when leaves are in excess. But, they always mix mulberry twigs with straw. Farmers also use the mulberry twigs as their fuel source after pruning of the garden. Left over twigs are allowed to dry in the garden itself. Refuses of rearing are also converted to valuable Farm Yard Manure for mulberry garden by the farmers by putting it in a pit and keeping it for 4-5 months prior to its use.

As mulberry is mainly propagated by cuttings in tropics and sub-tropics, certain quantity of pruned branches can be used for preparation of cuttings and the remaining as fuel wood. One hectare of mulberry garden yields about 12.1 MT of mulberry sticks. The energy generated/ha (50% moisture loss) is 27830 Kcal (@ 4600 calories/kg of mulberry wood). Accordingly, mulberry can be exploited by raising it as "energy plantation" in cultivable/wasteland/low lying areas/canal bund/road side/fringe areas of the forest etc., under various afforestation, watershed development and soil conservation programmes.

Traditional uses of Mulberry

The various species of the genus *Morus* viz., *M. laevigata*, *M. serrata*, *M. alba* and *M.indica* has varied uses which are enumerated below:

M. laevigata: The trees of this species produce sweet fruits which are used in juice and jam making in central India. In North-east India the wood is utilized as firewood, house building, furniture making, for making stocks, spokes, poles, shafts of carriages and casts. The wood is suitable for plywood making and paneling, carving and making of toys and tea chests. It is used for making tennis rackets. The straight log of the tree is used as support in house building work.

M. serrata: The wood is used for furniture making and carving, toys making, sports goods, agricultural implements and cheap types of rifles and guns.

M. alba: It is cultivated in the hilly and plain areas of India (Himalayan region) for silkworm rearing. It is also used as a avenue tree and in social forestry. Fruits are made into juice, liquor and stews. The wood finds use in sports goods industry. It is also used for house building, agricultural implements, furniture, for making spokes, poles, shafts and bent parts of carriages and carts. The stem bark is used for paper making.

M. indica: The cultivated forms belong to *M. indica* which are utilized in silkworm rearing. There are few profuse fruiting varieties occurring in Maharastra and Meghalaya which can be utilized as female parent in breeding programmes. The fruits are used for jam, jelly and juice making in Maharashtra. The pruned branches are used as fuel.

Medicinal Uses

The various parts of mulberry plant finds use in ayurvedic preparations. The leaves have diaphoretic and emollient effects. The leaves are used for making a decoction which can be used as a gargle to get relief from throat inflammation. The fruits are used to treat sore throat, depression, high fever and is both a coolant and laxative. The roots extract have hypoglycemic properties. The root bark is used as an anthelmintic, purgative and vermifuge. Mulberry root juice is administered to patients with high blood pressure. The chinese use the leaf tips from young leaves to boil with tea and use it to control blood pressure. The milky latex is used as a plaster for sores and for preparation of dermal creams.

Fruiting Habit

In South India, fruits are observed in two seasons in a year once during October - November and the other during March - May. However, whenever mulberry is pruned or defoliate flowering takes place along with sprouting of auxiliary buds followed by fruit formation. This feature of mulberry is utilized in mulberry breeding programmes. The immature fruits are green in colour but changes to purplish to violet black. In certain species like *Morus alba* the fruits are white to pinkish and very sweet. In *Morus lavigata* the fruits are very long sometime measuring up to 18cm.

Chemical Composition of Leaf

Chemical composition of leaf varies with variety and maturity. However on the basis of the analysis carried out at CSR& TI Mysore , the chemical composition of the leaf is as follows:

Leaf component	Range
Moisture	65 - 78 %
Protein	19 – 25 %
Minerals	10 – 15 %
Reducing sugars	1.2 – 1.9
Sugars	10 – 15 %

Pests and Diseases

Key pests of mulberry:

— *Maconellicoccus hirsutus* (mealy bug) – causing tukra in mulberry.

— *Diaphania pulverulentalis* – Leaf roller.

— *Spilarctia obliqua* – Bihar hairy caterpillar (Sporadic pest).

Minor pests of mulberry:

Thrips, Jassids, Scale insects, short horned grasshopper.

Tolerance of varieties to pests:

Main diseases of mulberry:

FOLIAR	Leaf spot
	Leaf rust
	Powdery mildew
	Leaf blight
	Bacterial blight
SOIL BORN	Root rot
	Root knot
	Nursery diseases (Stem-canker, cutting rot, collar rot and die-back)

The tukra incidence in rainfed areas was found to be maximum in S-34 variety followed by MR-2, BERC 776, MS7 and S-13.

The spread of tukra in V-1 variety is less compared to other varieties viz., Local, K-2, S-13, S-34, S-36 suggesting V-1 variety to be relatively tolerant to tukra.

8

Mulberry Cultivation in Japan

Mulberry is distributed in a wide area of tropical, subtropical, temperate and sub-arctic zones. Most of mulberry varieties cultivated in Japan belong to *Morus bombysis* Koidz, *M. alba* and *M. latifolia* Poiret. Varieties belonging to *M. bombysis* are primarily cultivated in cold regions, such as Tohoku district. *M. latifolia* varieties are mainly cultivated in warm places, such as Kyushu district. However, varieties of *M. alba* are cultivated in a wide area, from Kyushu district to Tohoku district, since it has middle traits of these two species. Besides, *M. acidosa* Griff. grows naturally and is cultivated in south-west islands including Okinawa islands. *M. kagayamae* Koidz. and *M. boninensis* Koidz. are indigenous to the remote islands of Hachijoujima and Ogasawara, respectively.

Ploidy of Mulberry

Generally, mulberry is a diploid plant with 28 chromosomes (2n=28). However, it is rich in ploidy and a lot of triploid varieties have been found especially among *Morus bombysis* Koidz. It is said that *M. cathayana* Hemsl. has tetraploid, pentaploid and hexaploid varieties. Both

M. serrata Roxb., indigenous to India, and *M. tiliaefolia* Makino, originally from Japan and Korea, are known to be hexaploid. *M. boninensis* Koidz. is a tetraploid being endangered due to cross contamination with *M. acidosa* Griff. *M. nigra* L. is dexoploid (2n=308), the largest number of chromosomes among phanerogams.

Inflorescence of ulberry

Mulberry has different varieties with male, female or androgynous flowers. Among the varieties with androgynous flowers, there are predominantly staminated, predominantly pistillated and even hermaphrodite types. Mulberry plants having anemophilous flowers cross each other easily and naturally.

Genetic Resources

The National Institute of Sericultural and Entomological Science (NISES) of Japan has collected and maintained 1,300 accessions of both indigenous and exotic origin. In order to utilize these genetic resources efficiently in breeding programs, morphological traits necessary for classification and identification, agronomic traits responsible for the resistance against diseases and pests, and commercial traits related to productivity and quality, are investigated. The information on these traits is very useful for achieving breeding targets.

Mulberry Breeding Method

It takes many years (approximately 15 – 20) to develop a new variety because mulberry is a perennial woody plant. Breeding targets should be set with a long-term view. So far, breeding targets have been such as high yield, high nutritional value and resistance against diseases and pests. But nowadays, new targets have been added to cope with changes in the sericultural system such as large numbers of silkworm reared and adaptability to densely planted fields suitable for mechanical harvesting.

Crossing is the major breeding method adopted for the development of new mulberry varieties. In this method, the choice of parent selection plays a vital role of in achieving the objective. For example, the variety "Shin-ichinose" was selected from the F1 generation of "Ichinose" x "Kokuso 21". This combination aimed to develop a variety showing high quality, high yield and resistance against lodging, considering that "Ichinose" shows good quality and high yield and that "Kokuso 21" has vigorous growth and an erect tree type.

Two other varieties, "Tokiyutaka" and "Oyutaka" were also developed and released from this crossing. Besides, "Ichinose" and "Kokuso 21" have been used as parents for other varieties. One of the major reasons why "Ichinose" and "Kokuso 21" have been selected as parents is that "Ichinose" is female and "Kokuso 21" is male. They have desirable traits, and crossing is easy and simple.

A lot of indigenous, natural triploid varieties, such as "Ichibei", "Fukushima Oha", "Akagi" and "Tagowase" have been distributed, especially in Tohoku area, northern part of Japan. Since the 1960's, polyploidy breeding has been produced artificially by colchicine, which is capable of doubling chromosome number. The significance of polyploidy breeding is that 1) the leaves of triploids are larger than diploids and the yield is expected to be higher, 2) crossing between different ploids accumulates more genes than crossing between diploids, which is expected to yield more heterotic effect, and 3) triploids show good leaf quality and resistant against coldness. Up to date, five triploid varieties, "Shinkenmochi", "Aobanezumi", "Mitsushigeri", "Yukimasari" and "Yukiasahi" have been developed and released using the polyploid breeding method.

Mulberry Breeding System

In Japan, mulberry breeding started at the government Sericultural Experiment Station in 1916, and two varieties, "Kokuso 13" and "Kokuso 70" were released to farmers in 1922. After that, with the wave of post-war economic recovery planning, three varieties, "Kokuso 20", "Kokuso 21" and "Kokuso 27" were released in 1949. However, these varieties were not popularized very much, despite their high leaf productivity, because they were rather sensitive to dwarf disease, which causes the most serious growth damage, and of slightly poor quality.

In 1958, a field trial was initiated in four, climatically different, regions (cold, snowy, temperate and warm) and 13 prefecture sericulture experiment stations. In 1971, a dwarf disease resistance test was added to the program. When a selected strain is officially recognized to have excellent characteristics, it is commercially released as a new variety. Up to now, 19 varieties have been released, including "Shinichinose", "Yukishinogi", "Minamisakari", "Shinkenmochi", "Hayatesakari", "Aobanezumi", "Mitsuminami", and "Senshin". Such a system is believed to be indispensable to allow the Japan's breeding network to release a new mulberry variety to farmers.

Mulberry varieties

Generally speaking, varieties belonging to *M. bombysis*, have lobed and shallow bottom leaves. Branches are brown or gray, and the winter buds are oval and sharp-pointed. Varieties of *M. alba* have lobed/unlobed leaves with whitish-gray or grayish-brown branches. *M. latifolia* varieties have large, unlobed, lustrous leaves and greenish-gray or whitish-gray branches. "Ichinose" and "Kairyo-nezumigaeshi" most widely cultivated in Japan, are *M. alba* varieties, and "Kenmochi", intensively cultivated in cold areas, is a variety of *M. bombysis*.

As mentioned above, 19 varieties were registered and released from 1971 to 1998 by the Ministry of Agriculture, Forestry and Fisheries (MAFF). Out of them, six varieties, "Minamisakari", "Hayatesakari", "Mitsuminami", "Hinosakari", "Mitsusakari" and "Senshin", are suitable for warm areas such as the Kyushu and Shikoku regions. Four varieties, "Shin-ichinose", "Tokiyutaka", "Oyutaka" and "Tachimidori", are suitable for temperate areas, such as the Kanto region. Five varieties, "Shin-kenmochi", "Aobanezumi", "Mitsushigeri", "Hachinose" and "Waseyutaka", are adaptable to cold areas, such as the Tohoku region. Four varieties, "Yukishinogi", "Yukishirazu", "Yukimasari" and "Yukiasahi", are adaptable to snowy areas, such as Niigata prefecture. There are some varieties, like "Hayatesakari", which have been dispersed into the Tohoku from the Kyushu area, which was its place of origin.

Physiology and Cultivation

Mulberry Leaf Production

The growth of mulberry is generally divided into three stages; new shoot development, growth and storage. New shoot development is a stage in which new shoots develop using reserves stored in stump or root the previous year. Growth stage is when carbohydrates and other substances are produced for vegetative growth by means of photosynthesis in leaves. Storage stage is when most photosynthetic substances are stored for next year's growth. In cultivated mulberry plants, however, photosynthetic organs are removed/harvested by pruning and leaf picking disrupting the growth stage. The exploited mulberry plants resume growth using the remaining storage substances.

Nutritive Value

Table 1. Amino acid content in mulberry leaf and minimum requirement for silkworm (mg/g DM)

Amino acid	Content	(%)	S.D.	C.V.	Minimum requirement**
Asp	20.49	(10.0)	3.63	17.72	
Thr	10.52	(5.2)	1.75	16.63	7
Ser	10.12	(5.0)	1.60	15.79	
Glu	23.23	(11.3)	3.96	17.03	
Pro	10.93	(5.4)	3.73	34.10	
Gly	12.02	(5.9)	1.95	16.22	
Ala	15.75	(7.7)	2.90	18.44	
Val	12.83	(6.3)	2.17	16.92	8
Cys	1.17	(0.6)	0.25	21.72	
Met	2.99	(1.5)	0.61	20.48	4
Ileu	10.04	(4.9)	1.88	18.68	8
Leu	19.45	(9.5)	3.10	15.93	8
Tyr	7.40	(3.6)	1.39	18.74	
Phe	12.26	(6.0)	2.06	16.78	8
GABA	2.26	(1.1)	0.69	30.70	
NH3	2.89	(1.4)	0.54	18.70	
Lys	12.33	(6.0)	2.58	20.91	8
His	4.61	(2.3)	0.82	17.78	5
Arg	12.96	(6.3)	2.72	20.95	8
Total	204.25	(100.0)			
N(%)	4.36		0.42	9.63	

* Mean value of 119 varieties tested.

**Minimum quantity required for silkworm growth.

The silkworm eats only mulberry leaves to make its cocoon, producing the silk. Mulberry leaves are rich in protein and amino acids (Table 1).

It is known that there is high correlation between leaf protein level and production efficiency of cocoon shell, which means cocoon shell weight to the total amount of mulberry leaves consumed by the silkworm (Figure 1.). Therefore, an increase in protein level of mulberry leaves may lead to improvements in cocoon productivity.

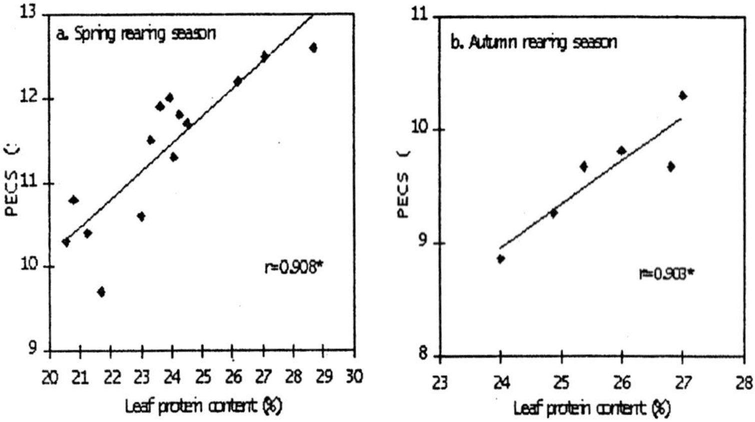

Figure 1. Correlation between leaf protein content and production efficiency of cocoon shell (PECS) in Mulberry varieties

Cultivation

According to the 1996 statistical data of Japanese sericulture, the area of mulberry gardens was 14,884 ha, including 1,172 ha of densely planted fields. Regarding planting density of mulberry gardens, normal planting with 600 - 1,000 plants per 10a is common. In case of densely planted field, aiming at early high yield and machinery harvesting, more than 2,500 plants/10ha are used.

Standard application of chemical fertilizer to mulberry garden is 30kg of N, 14kg of phosphate and 12kgof potassium/10a for alluvial soil, and 30kg of N, 16kg of phosphate and 20kg of potassium/10a for volcanic ash soil. In either case, application of at least 1,500kg of compost per 10a is recommended.

Training and harvesting

There are various training forms in mulberry cultivation according to various purposes. Maintenance of stump height is one of the typical forms of training. Based on the height from the soil surface: low cut (at the height of 15 – 30cm from soil surface), medium low cut (30 – 50cm) and medium cut (50 – 100cm) are under practice. The second one is a fist shape training method: if the plant is pruned at the fold every time, from where the shoots emerge, a fist is formed.

Meanwhile, if it is pruned slightly above it, leaving a definite bud, the plant height goes up gradually and a fist is not formed. This is called a non-fist shape training method. The third one is a lateral branch training method developed in the Yamanashi Prefecture: branches adjacent to plants within a row are held down and tied up with a wire and the shoots emerge from the buds of the lying branches.

Harvesting methods vary with the rearing scale and frequency. Basically there are two methods, spring pruning (for summer-autumn rearing season) and summer pruning (for both spring rearing and late autumn rearing seasons). There are also circle harvesting method (spring pruning and summer pruning alternatively every year) and alternative harvesting method (alternating spring and summer pruning to half of the same plant). These two methods are adopted to secure enough yield by sustaining the tree vigor. Meanwhile, in densely planted fields, mechanical harvesting is so essential that low pruning, at a point near the ground to prevent stump formation, is desirable.

Propagation

Mulberry propagation is generally carried out by grafting and by cutting methods. Root grafting prevails because it is easy to handle and the grafted saplings have a high survivability. The cutting method can be with hard wood (using the branches grown in the previous year) and soft wood (using the spring sprouted shoots). In case of mulberry varieties of poor rooting ability, treatment with plant hormone is advised to stimulate rooting. Recently, tissue culture derived saplings are also being produced.

Other Utilization

Mulberry was originally cultivated in Japan, as well as in other countries, for sericulture. Recently, however, mulberry has being re-evaluated due to its functional characteristics and being utilized for various purposes:

— *Fruit*. Mulberry fruit changes the color from green to black purple thmes and especially the root bark has been used as an herbal medicine due to its properties to reduce high blood pressure. Mulberry leaf is rich in gamma-aminobutylic acid, effective against high blood pressure, and in alanine, effective against hangover. Moreover, it was found that deoxynojirimycin, which is said to have an effect to lower the blood-sugar level closely related to diabetes, is abundant in mulberry leaf. That is why nowadays, mulberry tea is considered as a health food.

— *Paper production.* Mulberry grows faster than other woody plants and is said to be suitable for high biomass production. Mulberry branches are being used as raw material for paper production.

— *Mushroom production.* Mulberry stem and stem powder are found to be good source of media for mushroom production.

— *Animal feed.* The use of mulberry for animal production in Japan, is being reported in other articles on this E-conference.

9

Mulberry Cultivation in Brazil

Mulberry (*Morus* sp.), originally from Asia, was introduced in Brazil during the colonial times. Nowadays, Mulberry is widely known, frequent in orchards, recreational rural households and gardens. Its fruits are very well appreciated for direct consumption and for making marmalades. However, commercial mulberry cultivation occurs only in regions where it is associated with sericulture.

Table 1. Areas with mulberry

State	Mulberry area (ha)
Goiás (GO)	24
Mato Grosso do Sul (MS)	459
Minas Gerais (MG)	77
Paraná (PR)	32,428
Santa Catarina (SC)	114
São Paulo (SP)	4,645

According with the latest statistics, the mulberry area in Brazil covers approximately 38 thousand hectares (Table 1), state of Paraná being the largest concentration of that area with 32.4 thousand hectares (Table 2) followed by western portion of state of São Paulo, with 4.6 thousand ha (Table 3) and by smaller areas in states of Goiás, Mato Grosso do Sul, Minas Gerais and Santa Catarina.

Table 2. Areas with mulberry in Paraná State

Counties	Area (ha)
Apucarana	610,00
Campo Mourão	1243,76
Cascavel	1929,52
Cornélio Procópio	90,59
Curitiba	151,01
Francisco Beltrão	662,38
Guarapuava	213,68
Irati	1,16
Ivaiporã	2240,59
Londrina	630,44
Maringá	8775,97
Paranavai	4600,93
Pato Branco	501,72
Ponta Grossa	414,69
Santo Antônio Da Platina	1970,38
Toledo	762,87
Umuarama	7625,96
Total	32425,76

Table 3. Area with mulberry in São Paolo State

Regional agricultural divisions	Area (ha)
Andradina	243,04
Araçatuba	26,88
Araraquara	66,51
Assis	125,89
Avaré	21,28
Bauru	554,13
Catanduva	78,40
Dracena	236,32

General Salgado	446,68
Itapetininga	5,15
Itapeva	41,89
Jaboticabal	2,24
Jales	35,84
Jaú	43,30
Limeira	18,37
Lins	898,80
Marília	318,86
Ourinhos	161,62
Piracicaba	15,68
Presidente Prudente	221,85
Presidente Venceslau	49,05
São João da Boa Vista	34,99
São José do Rio Preto	193,76
Tupã	775,04
Votuporânga	29,68
Estado de São Paulo	4645,25

It must be highlighted that all production of these areas is for the silkworm, and no surplus is available for other uses. Plant management includes frequent pruning which prevents plants from blossoming and fruit development, thus, plant sex is irrelevant.

According Fonseca and Fonseca, the first attempt to commercially cultivate mulberry occurred in the Rio de Janeiro State, as initiative of the Emperor D. Pedro II, in the middle of 19th century. After a period of decadence, it was resumed only in 1923, in the eastern parts of São Paulo State following the Italian immigration. Later eastern São Paulo showed a decline in sericulture and western São Paulo flourished, coinciding with Japanese immigration. Nowadays. sericulture is concentrated in the northern part of Paraná State.

The causes that determined the movements of mulberry cultivation are not clear, but they must be related to various factors such as: the low income condition prevailing among rural population who sees sericulture as secure source for income generation; with the appearance of other economic alternatives; and with soil exhaustion, production and profitability decline, resulting in sericulture opening space to other activities with more intensive in capital requirements.

Mulberry Varieties and Clones

In Brazil, the totality of cultivated mulberry varieties appear to belong to *Morus alba.* In São Paulo State, there is an active germplasm bank (BAG) at the Estação Experimental de Zootecnia (Livestock Experimental Station) in Gália county; one collection centre at the University of São Paulo state (UNESP) in Jaboticabal county; one in a spinning mill (under the trade name of *Fiações de Seda Bratac*), in Bastos county; and one at the Agronomic Institute of Paraná (IAPAR), in Londrina.

The Estação Experimental de Zootecnia has sericulture research as mandate and its BAG includes three collections: a) variety collection (32 clones), b) the Instituto de Zootecnia collection (42 clones), and, c) the Fukashi-Miura collection (14 clones).

Little information is available regarding which clones are used in the 38,000ha planted with mulberry. Commercial companies have disseminated the Miura and Korin clones, due to the great supply of planting material. These clones together with the Calabresa variety make the large majority of cultivated area.

Genetically modified clones (IZ & FM) are very productive and with more nutritious leaves, but being destined completely to silkworm feeding, their expansion has been limited.

Production data for this report were obtained from research carried out at Instituto de Zootecnia. Considering that the purpose of the work was fresh biomass production for silkworm feeding, the data are in fresh leaf weight. However, according with Almeida *et al.*, it is known that the leaf:stem ratio is 1:1 when the cut is made at 90d. For more frequent cutting leaf:stem ratio can vary.

Variety Collection

1. *Branca da Espanha* (Spanish white). Imported from Spain. It shows good development, good adaptation, it is productive and precocious. Good propagation through cuttings.
2. *Calabresa*—Imported from Italy. Optimal adaptation. It is productive, rustic and very precocious. Easy propagation through cuttings. Estimated production is 5,079 kg/ha/year.
3. *Catânia 1.* Imported from Italy. Well adapted, is vigorous and very productive. Not well propagated through cuttings.
4. *Catânia 2.* Imported from Italy. Well adapted, vigorous and highly productive. Not well propagated through cuttings.
5. *Catânia Paulista.* From the extinct Livestock Experimental Station at Limeira, São Paulo State. Its characteristics have some resemblance to two Catânia varieties above. It is precocious, productive and vigorous. Not well propagated through cuttings.
6. *Contadini.* Imported from Italy. Well adapted, productive and precocious. Very good propagation through cuttings.

7. *Fernão Dias*. From Fernão Dias county of São Paulo State. It is precocious and productive. Good propagation through cuttings.
8. *Flório*. Italian origin. Not very well adapted. It is precocious but not productive. Not well propagated through cuttings.
9. *Formosa*. Originary from Taiwan. Very well adapted, productive, precocious and vigorous. Easily propagated through cuttings. Estimated production is 8,608 kg/ha/year.
10. *Galiana*. Originary from Livestock Experimental Station at Gália, São Paolo State. Medium tardy, vigorous and rustic. Not well propagated through cuttings. Data for 1m branch: leaf number, 22; mean leaf weight, 5.86g; mean stem weight, 53.1g; leaf weight:stem weight, 2.43; internodal distance 4.5cm.
11. *Iamada*. Originary from Promissão county, São Paolo State. It is precocious, but little productive. Good propagation through cuttings.
12. *Kokuso 21*. Imported from Japan. Not well adapted, is tardy, produces few branches that grown slowly. No propagation through cuttings.
13. *Kokuso 27*. Imported from Japan. Not well adapted, tardy, produces few branches that grown slowly. No propagation through cuttings.
14. *Lopes Lins*. Originated from Tietê county, São Paolo State. It is precocious and productive. Good propagation through cuttings.
15. *Miura*. Originated from Bastos county, São Paolo State. It has some similar characteristics to the Calabrese variety. It is precocious, rustic and productive.

16. *Moretiana*. Imported from Italy. Very good adaptation, productive and rustic. It is little tardy, no propagation through cuttings.

17. *Moscatela*. Originated from Italy. Good adaptation, prococious, rustic and productive. Good propagation through cuttings.

18. *Nezumigaeshi*. Imported from Japan. Good adaptation, productive, rustic and vigorous. No well propagated through cuttings.

19. *Nostrana*. From the extinct Livestock Experimental Station at Limeira, São Paolo State. It is rustic, precocious and vigorous. Easy propagation through cuttings.

20. *Paduana*. Coming from Borborema county, São Paolo State. It is precocious, rustic and productive. Good propagation through cuttings.

21. *Pêndula*. Originated from Rio de Janeiro State. It is precocious, rustic but little productive

22. *Rosa*. Originated from Italy. Good adaptation but low productivity. Slow development. Good propagation through cuttings.

23. *Rosa da Lombardia* – Originary from Italy. Bad adaptation, it medium tardy, low productivity. Slow development and precocious leaf maturation. No good propagation through cuttings.

24. *Rosol*. Originated from Registro county, São Paolo. It is precocious, rustic, vigorous and productive. but leaves are very wrinkled, coarse and easily broken, little adapted to feeding the silkworm.

25. *Selvagem*. From the extinct Sericulture Service at Campinas, São Paolo State. It is rustic, vigorous and precocious, showing low productivity, with

excessively multiblobed leaves with reduced useful foliar area.

26. *Serra-das-Araras*. Coming from Ararasmountain range situated in Rio de Janeiro State. It is very precocious, rustic and with low production. Very intense blossoming Easy propagation through cuttings.
27. *Siciliana*. Coming from Barbacena county, Minas Gerais State, It is precocious, rustic but with low production. Easily propagated through cuttings.
28. *Talo Roxo*. From the extinct Sericulture Service in Campinas, São Paolo. It is precocious, with low production, and easy propagation through cuttings.
29. *Tietê*. Coming from Tietê county, São Paolo State. It is precocious, rustic and with low production. Good propagation through cuttings.
30. *Ungaresa*—From the extinct Livestock Experimental Station at Limeira, São Paolo State. It is rustic, precocious and productive. Leaves are coarse, no appreciated by silkworm.
31. *Korin*. From Fiações de Seda Bratac collection, is very vigorous and productive. Good propagation through cuttings.

The production and agrobotanic characteristics of the varieties above are contained in Tables 4 and 5.

IZ Collection

This is a collection made by selected clones, originary from breeding programs conducted by researchers from Livestock Institute (Instituto de Zootecnia or IZ), which belongs to the Secretary of Agriculture of São Paulo State

Table 4. Quantitative Characteristics of some mulberry varieties (data for 1m-branch)

Variety	Leaf number	Mean leaf weight (g)	Mean branch weight (g)	Leaf/ Branch ratio	Internodal length (cm)
01. Branca da Espanha	24	4.65	42.5	2.63	4.2
02. Calabresa	26	3.04	33.5	2.36	3.8
03. Catânia 1	26	5.79	64.3	2.35	3.8
04. Catânia 2	26	6.09	50.9	3.10	3.8
05. Catânia Paulista	25	5.44	61.8	2.20	4.0
06. Contadini	31	3.29	38	2.69	3.2
07. Fernão Dias	25	4.10	38.6	2.67	4.0
08. Flório	25	2.40	37.5	1.60	4.0
09. Formosa	24	4.66	41.5	2.70	4.2
10. Galiana	22	5.86	53.1	2.43	4.5
11. Iamada	30	2.52	52.7	1.14	2.8
12. Kokuzo 21	31	7.29	75.1	3.01	3.2
13. Kokuzo 27	30	7.88	75.0	3.15	3.3
14. Lopes Lins	19	5.30	41.9	2.41	5.2
15. Miura	24	4.24	38.4	2.66	4.2
16. Moretiana	33	2.67	46.8	1.88	3.3
17. Moscatela	24	3.91	31	3.04	4.2
18. Nezumigaeshi	47	3.54	65.1	2.55	2.1
19. Nostrana	21	4.23	30.9	2.88	4.8
20. Paduana	20	8.04	44.6	3.61	5.0
21. Pendula	24	3.48	32.2	2.61	4.2
22. Rosa	28	2.47	35.6	1.91	3.6
23. Rosada Lombardia	31	4.29	62.2	2.14	3.2
24. Rosol	16	9.05	44.1	3.29	6.2
25. Selvagem	21	2.68	25.0	2.24	4.8
26. Serra das Araras	42	1.98	34.5	2.44	2.4
27. Sicicliana	23	3.17	29.5	2.48	4.3
28. Talo Roxo	24	3.95	40.1	2.37	4.2
29. Tiete	21	4.75	43.1	2.32	4.8
30. Ungareza	26	2.95	32.3	2.39	2.8
31. Korin	-	-	-	-	-

Table 5. Agrobotanic characteristics of same mulberry varieties

	Size (cm^2)	Shape	Base	Edge[1]	Surface	Bark color	Bud[2]	Flower	Size[2] Colour
B. da Espanha	380	Entire/Lobed	Truncated	S	Smooth/Glossy	Greyish Yellow-	S	?	S/White
Calabresa	240	Five lobes	Lobed	D	Smooth/Glossy	Dark greyish brown	S	?	M/Purple
Catânia 1	330	Entire/Ovate	Truncated	D	Undulated/ Glossy	Light greyish yellow	S	?	L/Purple
Catânia 2	335	Entire/Ovate	Truncated	D	Undulated/ Glossy	Light brown	M	?	L/White
Catânia Paulista	252	Entire/Ovate	Rounded	D	Undulated/ Glossy	Whitish brown	L	?	-
Contadini	257	Entire/Ovate	Rounded	D	Undulated/ Glossy	Greyish brown	M	?	M/Purple
Fernão Dias	274	Entire/Ovate /Lobed	Cordated	D	Smooth/Glossy	Whitish brown	M	?	M/Purple
Flório	150	Entire/ Cordiform	Cordated	S	Smooth/Glossy	Brown	S	?	S/Purple
Formosa	339	Entire/Ovate /Lobed	Cordated	S	Smooth/Glossy	Dark brown	M	?	M/Purple
Galiana	405	Lobed	Truncated lobed	S	Smooth/Glossy	Greyish brown	L	?	M/Purple
Iamada	184	Entire/Ovate /Lobed	Cordated lobed	D	Smooth/Glossy	Greyish brown	M	?	M/Purple

Contd....

Contd....

Kokuzo 21	408	Entire/Ovate	Linear Round	C	Smooth/Glossy	Light greyish yellow	M	?	-
Kokuzo 27	405	Intire/Ovate	Round	C	Smooth/Glossy	Light greyish yellow	S	?	-
Lopes Lins	330	Entire/Lobed	Cordated	C	Smooth/Glossy	Light greyish brown	S	?	S/Purple
Miura	276	Entire/Lobed	Cordated Lobed	D	Smooth/Glossy	Light greyish brown	L	?	M/Purple
Moretiana	181	Entire/ Cordiform	Cordated	S	Smooth/Glossys	Brown	M	?	S/Purple
Moscatela	282	Entire/Ovate	Truncated	S	Smooth/Glossy	Light brown greyish	L	?	S/Purple
Nezumigaeshi	213	Five lobed*	Truncated Lobed	S	Smooth/Glossy	Light yellow greyish	S	?	-
Nostrana	250	Entire/Ovate	Rounded	D	Undulated/ Glossy	Dark brown	M	?	M/Purple
Paduana	404	Entire/Ovate /Lobed	Cordated	D	Undulated/ Opaque	Greyish brown	L	?	-
Pendula	232	Entire/Ovate /Lobed	Truncated Linear	D	Smooth/Glossy	Greyish brown	L	?	M/Purple
Rosa	177	Entire/ Cordiform	Cordated	D	Smooth/Opaque	Brown	M	?	S/Purple
R.da Lombardia	289	Entire/ Cordiform	Deeply cordated	S	Smooth/Glossy	Light brown	S	?	M/Purple
Rosol	500	Entire/Ovate	Linear	D	Blistered	Whitish brown	S	?	S/Purple

Contd....

Contd....

Selvagem	210	Lobed	Lobed	S/D	Smooth/Glossy	Dark brown	L	?	M/Purple
Serra das Araras	144	Entire/Ovate	Cordated	S	Smooth/Glossy	Brown	L	?	S/Purple
Sicicliana	253	Entire/Ovate /Lobed	Truncated	S/D	Undulated/ Glossy	Dark brown	L	?	-
Talo Roxo	282	Entire/ Cordiform	Cordated	D	Undulated/ Opaque	Dark reddish brown	L	?	L/Purple
Tiete	272	Entire/Ovate /Lobed	Truncated	D	Smooth/Glossy	Brown	M	?	M/Purple
Ungareza	209	Entire/Ovate	Truncated Linear	S/D	Smooth/Opaque	Dark brown	S	?	-
Korin --	-	-	-	-	-	-	-		

1 S = serrated; D = dentated; C = crenated

2 S = small; M = medium; L = large

(Table 6). The production and agrobotanic characteristics of the IZ clones are shown in Tables 7.

Table 6. Origin and estimated production of IZ collection clones

Clone	Cross-breeding	Estimated production (kg/ ha/year)
IZ30	Random breeding	8,559
IZ40	Random breeding	9,931
IZ64	Random breeding	7,945
IZ1/17	Fernão Dias x Catânia Paulista	7,686
IZ2/2	Calabresa x Catânia Paulista	6,197
IZ3/2—Issaokina	Contadini x Catânia Paulista	8,597
IZ5/2—Capucho	Branca de Espanha x Catânia	7,895
IZ6/2	Lopes Lins x Catânia Paulista	5,438
IZ10/1—Campinas	Lopes Lins x Catânia Paulista	7,984
IZ10/4	Lopes Lins x Catânia Paulista	6,698
IZ12/3	Fernão Dias x Catânia Paulista	6,104
IZ13/6—Luiz Paolieri	Fernão Dias x Kokuzo	11,844
IZ15/1	Calabresa x Nezumigaeshi	6,304
IZ15/7—Rio da Pedras	Calabresa x Nezumigaeshi	6,501
IZ19/1	Talo Roxo x Kokuso 27	5,119
IZ19/13—Rosa da Fonseca	Talo Roxo x Kokuzo 27	10,177
IZ23/8	Rosol x Catânia Paulista	5,555
IZ29/1—Sempre Verde	Capinas x Nezumigaeshi	8,224
IZ42/12	Catânia x Siciliana	5,001
IZ51/1	Brabca de Espanha x Nezumigaeshi	7.878
IZ56/4 0 Tamarina	Formosa x Catania Paulista	12,043
IZ57/2—Javanesa	Fromosa x Kokuso 27	9,353
IZ1/1	Fernão Dias x Catânia Paulista	
IZ1/2	"	
IZ1/3	"	
IZ1/12	"	

IZ1/16		
IZ2/1	Calabresa x Catânia Paulista	
IZ3/1	Contadini x Catânia Paulista	
IZ4/1	Moretiana x Catânia Paulista	
IZ4/4	"	
IZ5/1	Calabresa x Catânia Paulista	
IZ6/1	"	
IZ6/3	Lopes Lins x Catânia Paulista	
IZ6/7	"	
IZ11/9	Formosa x Kokuso 21	4,526
IZ14/1	Branca de Espanha x Catânia Paulista	
IZ16/3	Fernão Dias x Catânia Paulista	5,310
IZ18/19	Catânia x Catânia Paulista	
IZ23/3	Rosol x Catânia Paulista	4,049
IZ23/3	"	

Table 7. Quantitative characteristics of mulberry clones of the IZ collection(data per 1m-branch)

Clone	Leaf number	Mean leaf weight (g)	Mean branch weight (g)	Leaf/ branch ratio	Internodal length (cm)
IZ 30 -	27	2.52	35.0	1.70	3.71
IZ 40	34	2.35	49.0	2.35	2.95
IZ 64	26	4.70	38.0	3.00	3.85
IZ 1/17	19	5.28	48.9	2.05	5.3
IZ 2/2	24	3.06	49.9	1.47	4.2
IZ 3/2	27	4.10	49.9	2.22	3.7
IZ 5/2	24	4.13	46.3	2.14	4.2
IZ 6/2	22	4.09	48.6	1.85	4.5
IZ 10/1	22	5.56	46.8	2.62	4.5
IZ 10/4	23	4.26	53.7	1.82	4.3
IZ 10/8	19	4.27	38.0	2.14	5.3
IZ 12/3	25	3.88	41.9	2.32	4.0
IZ 13/6	25	4.82	53.0	2.28	4.0
IZ 15/1	25	3.74	52.1	1.80	4.0
IZ 15/7	26	1.83	34.6	1.83	3.8
IZ 19/1	28	3.10	50.7	1.71	3.6

IZ 19/13	32	4.06	54.9	2.37	3.1
IZ 23/8	23	5.62	52.9	2.44	4.3
IZ 29/1	28	3.57	48.0	2.08	3.6
IZ 42/12	22	3.34	37.4	1.97	4.5
IZ 51/1	21	3.89	35.5	2.30	4.8
IZ 56/4	23	4.24	49.4	1.97	4.3
IZ 57/2	25	4.16	44.2	2.35	4.4

FM Collection

Mulberry clones selected by Mr. Fukashi Miura (FM) from Fiações de Seda Shoei-Bratac, São José do Rio Preto, São Paolo State:

— *Shima Korin 01—SK02—SK03—SK04* - From cross-breeding Shimagoa and Korin varieties.

— *Shima Miura 11—SM12—SM13—SM14—SM15* - From cross-breeding Shimagoa and Miura varieties.

— *Fukashi Miura 3/1—FM6/3—FM10—FM30—FM3/2* - From selection into plants origine by seed

Mulberry Field Management

Weather

Weather: Optimal temperature required for mulberry is situated between 24—28°C; below 13° growing is stopped and so is branching and bud formation. From August to May, mulberry vegetative growing is practically continuous.

Rainfall: Ideal rainfall for mulberry could be situated between 1000—1500mm, distributed along the productive period and this condition met in São Paulo e Paraná States.

Relative Humidity: In general, required air humidity for mulberry is between 65% - 80%. Higher precipitation and soil humidity are contributing factors for improved leaf quality.

Photoperiod: Major mulberry areas in Brazil are located between 20o e 30o South, and it seems that there is no photoperiodic limitations to plant growing.

Altitude : In São Paulo State, in general, mulberry crops are located in altitudes that varies from 300 - 700m, far from the limit tolerated by mulberry.

Drought resistance : No research data are available relating to drought resistance for varieties cultivated in Brazil. Seasonal supply of green leaves shown by some varieties even in drought periods was qualitatively observed for some clones, for example IZ 29/1.

Soil

Although mulberry requires adequate soil quality, it can grow in most of soils available in São Paulo State. However, shallow soils should be avoided, as well as compacted, hard, impermeable and swampy soil. Whenever possible, preferred soils are deep, fertile, well-drained, friable, porous, with adequate water retention capacity and a mild acidity (pH 6.2 - 6.8).

Fertilization

Organic (manure, organic composites etc): In general, poultry manure is applied, with a minimum dose of 0.5 kg per plant, during the winter time, at the depth of 20 to 30 cm.

Field experiments demonstrate higher productivity when mulch is applied over the mulberry plants. When available, its utilization is highly desirable, demanding, though, some necessary precaution. For examples, when residues of different origins with very high C/N ratio are used, it is recommended that chemical Nitrogen fertilizer should be applied jointly in order to accelerate the decomposition of organic material and avoid competition for Nitrogen in the soil.

Chemical Fertilizers

According to the literature there is no consensus on the ideal chemical fertilizer for mulberry plants. Among other factors, quantities to be applied will obviously depend on soil fertility.

Table 8. Typical composition of various organic materials from fresh animal and vegetal origins

Material	C/N	Humidity	C	N	P	K	Ca
			g/kg				
Fresh bovine manure	20	620	100	5	2,6	6	2
Composted bovine manure	21	340	320	15	12	21	20
Poultry Manure	10	550	140	14	8	7	23
Swine Manure	9	780	60	7	2	5	12
Castor oil seed meal	10	90	450	45	7	11	18
Mucuna	20	870	60	3	0,6	3	2
Crotalaria juncea	25	860	70	2,8	0,4	3	2

Material	Mg	S	Zn	Cu	Cd	Ni	Pb
		g/kg			mg/Kg		
Fresh bovine manure	1	1	33	6	0	2	2
Composted bovine manure	6	2	217	25	0	2	1
Poultry Manure	5	2	138	14	2	2	17
Swine Manure	3	-	242	264	0	2	3
Castor oil seed meal	5	-	128	73	-	-	-
Mucuna	0,4	-	6	3	-	-	-
Crotalaria juncea	0,4	-	2	1	-	-	-

According to Rùbia *et al.*, in cases of low soil contents in P and K and all silkworm residues being applied in mulberry cultivation, the following chemical prescription could be suggested (in kg/hectare/year): 250kg of N, 65kg of P_2O_5, and 156kg of K_2O.

In the case of manure utilization, the data presented in Table 8 can be of help to determine nutrient application rates.

Pests

Principal Pests

Pseudaulacaspis pentagona. This insect attacks mulberry branches and stems, introducing its mouth apparatus (stylus) into the sap plant conductors and sucking a great deal of the plant sap. These insects shield themselves under small structures and, when the attack is intense, those structures fully cover branches and stems of the plant. The plant is quite weakened by the attack, showing cracks that allow the entrance of pathogenic micro-organisms. In boggy soils, shadowed, poorly aired, or yet in case of crops that are not managed adequately, this attack is more severe.

Naupactus spp – These beetles belong to the Curculionidae family. Mulberry leaves feed adult phases of the insect, causing severe reduction in leaf surface, and root system feeds larvae phases. As a result, plant vigor decays, once nutrient absorption collapses and pathogenic micro-organisms are eased through plant injures. The *Naupactus versatilis* form has been more frequently detected in Paraná State. Its adult form presents a length ranging from 11 to 14mm, with metallic green colour, which becomes darker with age, due to the loss of scale-like structures.

Migdolus fryanus Westwood—Coleoptera from Cerambycidae family. This poliphagus beetle, biologically little known, has been causing economic damages in several crops, namely in sugar-cane, and more recently, in mulberry. Adult males have black coloration (in most of the cases), dark brown or red brown, and their total length size range from 12.1 and 37.0mm. Adult females, on the other hand, show a red-brownish coloration (in most of the cases) or dark brownish, and their total length ranges from 19.5 to 35mm. The male exemplars have membranous and functional wings, with antennas that reaches approximately half the body size. Female ones have reduced membranous wings, non-functional, and antennas much shorter than the males.

Larvae causes the totality of the damage to the mulberry plants. Those larvae meet in the root system the ideal source of nutrition, so destroying them. Insect dispersion happens in the larvae phases, because in this period it is observed an increased mobility in soil environment, establishing a net of channels that are likely to be used for adults in order to reach the soil surface during mating period.

Although most of the times plants are affected only partially, attacks can completely destroy the affected plants.

Regarding the population control of the insect, there are very few alternatives. due to biological and behavioral aspects. One of the obstacles is the fact that larvae lodge themselves in a depth of 3 to 4m into the soil, and eggs are laid in different depths.

Although there was a generalized belief that attacks were more frequent in sandy soils, nowadays studies on the subject have demonstrated that there is no linkage between the type of soil and the attack of *Migdolus*

"Cutting-prone" ants—Sauba ant – Ant hills are true subterranean urban structures, endowed of a great number of individuals (above tem million, in a adult ant hill), and that can reach till seven metres in depth, below soil line.

Quenquéns ants from the genus *Acromyrmex.*- They are smaller than sauba ants and, in general, ant hills from this genus hardly reach depths larger than 50 cm. The communication from one ant hill to another is made by *ocelli* (orifices on the soil surface). In an ant hill 3 or more years old, three types of *ocelli* are observed: one for transportation of land, one for ventilation, and one for food provision.

Nematode diseases—Meloidogyne spp—Those worm species attack mulberry producing root tumours (root-knots) that make it hard for the plant to absorb nutrients and water, which could cause severe damage to the crop. Plants attacked present reduction in size, with leaves appearing to lack water, becoming yellower and falling down, reducing sensibly the production of leaves per area. Mulberry vitality deteriorates gradually and, in the later stages, the plant dies.

Main Bacteria and Fungus Diseases

1. *White root rot,* caused by fungus from genus *Rosellinia.* This type of fungus live as saprophyte in decaying organic matter, mainly from plants left over the soil, and they can become pathogenic to cultivated plants. The diseased mulberry plants become weak, and root is led to be rotten, The *hyphae* emerging from the various fruiting bodies

attack the smaller roots of the healthy plants, and spread to the main root system causing white root rot.

2. *Violet root rot,* caused by *Helicobasidium mompa* Tanaka fungus. First symptoms from this disease appear when mulberry does not sprout in the beginning of spring or when leaves suddenly appear to wither in the middle of summertime. The main disease characteristic is the existence of similar velvet veilings that colors of dark-brownish the root bark. There is the appearance of several reddish filamentous substances that colors all the root bark that are the hyphes from pathogen.
3. *Mulberry Withering*—This disease has been its frequency verified in mulberry cultivations of State of São Paulo, mainly in the region of São José do Rio Preto, having as main characteristics the rottening of root. Several samples of attacked plants were examined in research institutes and up to the present time, non-conclusive diagnosis are available.
4. *Oídium – Phylactinia corylea Prest Karst.*—Oidium is a very common disease in mulberry plants and could cause severe damage when incident on leaves, making them unable to feed silkworm, once the fungus absorbs all the leaf nutrients.

   It frequently occurs in older leaves, that become covered by a grayish-white powder, made from pathogen structures, being more frequent during the spring and summer, periods when favorable conditions for its development are provided.
5. *Rusty spot Cylindrosporium mori Targioni Tozzeti*—In the beginning, the spots are smaller (less than 1

mm of diameter) and with the development of the process, they can reach more than 1 cm, with a brownish coloration, and yellowish edges. The spots use to appear in the superior part of the leaf, in the petiole and over the young branches.

In severe attacks, mulberry leaves drastically fall, causing a great deal of damage.

6. Other leaf spots caused by fungus—It may occur other spot attacks cause by fungus from *genera Cercospora* sp, *Alternaria* sp, *Helminthosporium* sp e *Colletotrichum* sp. Seasonally, they are more frequent during spring and summer, in periods of higher humidity and temperature.
7. *Bacteria spots—Bacterium mori, Bacillus curbonianus* (Macchiatti), *Bacterium moricolum.*—Mulberry is susceptible to this kind of spots, both at any age and in whatever aerial part of the plant. Seasonally, the disease is more frequently observed in summer (higher humidity and temperature).

On the leaves, initial symptoms appear as small and sparsely distributed in necrosed areas, that coalesce and torn apart in a posterior phases, making it possible the exudation of yellowish pus on the lesion. The shootings can also die, causing a oversprouting.

10

Utilisation of Mulberry

Because of its low economic return, production of green fodder on cultivated land is not very common in India. Furthermore, the lack of sufficient arable land with irrigation facilities also restricts fodder production. Such a situation is more acute in the hilly areas of the country. The huge cattle population in areas depends only on the leaves of certain tree species.

Trees have the advantage that they can be planted on hills, on wasteland and at the edges of ponds, and canals, etc. Trees can also be grown on boundaries where regular crops cannot be grown. They have the potential to produce as much green fodder per unit area as agricultural fodder crops. Trees do not need to be watered since they can extract water through their deep and extensive root system. The abilities of trees to extract water from deep underground layers and to withstand drought are outstanding advantages over agricultural fodder crops.

Trees also do not need such heavy inputs in the form of fertilisers, pesticides, fungicides and labour, as are needed for growing agricultural crops. Of all the tree species used as fodder, mulberry (*Morus* sp.) occupies a significant place as it grows anywhere, either in the form of have tree or shrub. Mulberry is believed to have originated in the foothills of the Himalayas and has been exploited for rearing silkworm in China since at least 3,500 BC. Sericulture then spread to India and to other countries. Although mulberry is the only food for the silkworm, its casual use as animal feed has also been known for a long time.

Mulberry Distribution

Mulberry is distributed throughout the world as it has enough plasticity to survive under disruptive environmental conditions. At least 30 countries are now producing raw silk with mulberry and silkworms. In some of these countries, the use of mulberry leaf as fodder has also been adopted. Mulberry can be grown successfully in all conditions, even in tropical, subtropical and temperate climates. It can survive with rainfall ranging from 400-4 500 mm per annum. Although the optimum temperature for growth is between 18 and 30°C, mulberry can survive even when the temperature goes beyond 48°C or below 0°C.

Mulberry can therefore be considered a universal plant, which can grow anywhere under varied climatic conditions. Although this does not guarantee successful silkworm rearing, the plant's wide adaptability can be exploited in plantations, fully or partly, for animal feed.

Mulberry as Fodder

The important role of green leaves in supplementing animal feed is unquestionable. In the developing countries, cereal straws and grasses are fed to animals, but they cannot support full performance because of their poor nutritive value. In most places where concentrates are also provided, the diet is not balanced. Mulberry leaf supplementation can improve the efficiency of the whole diet.

In addition, because of increasing pressure from the human population and to higher incomes earned from cereal and cash crops rather than from forage crops, more agricultural land cannot be set aside for fodder production. Another consequent advantage of mulberry is that it can be grown either as a tree or a shrub and harvested several times a year. In countries such as India, where mulberry is primarily grown for sericulture, excess leaves and leftovers are fed to cattle, sheep and goats. In hilly areas, where mulberry trees are abundant, their leaves are fed to animals.

Of the different species available, *Morus alba* and other species that are suitable for silkworm rearing are cultivated in the plains and on hilly ground. *Morus serrata, Morus laevigata* and *Morus australis* are grown in the hills. Because of their deep root system, the leaves remain green for most of the year, except in winter when the leaves fall. As a result of the above advantages, mulberry can be considered as a perennial source of feed for most of the year.

Mulberry fodder is considered to be of good quality. Leaf yield varies with nutrient supply and irrigation. In general, under irrigation with the recommended dose of fertiliser, mulberry yields nearly 35-45 tonnes of fresh leaf/ha/year. On a dry matter basis, leaves contain an average of 20-23 percent CP, 8-10 percent, total sugar, and 12-18 percent minerals. The leaves contain nearly 70 percent moisture. The cell wall constituents are neutral detergent fibre 45.6 percent, cell contents 54.4 percent, acid detergent fibre 35.0 percent, hemicellulose 10-40 percent, lignin 10 percent, cellulose 21.8 percent and silica 2.7 percent.

The contents of minerals are as follows: magnesium, 0.52-1.25 percent; chlorine, 0.02-0.29 percent; sulphur, 0.18-0.76 percent; potassium 0.93-3.19 percent and sodium 0.13-0.23 percent. Tannin accounts for 0.85 percent in leaf DM. The chemical composition of the leaves varies from season to season. In hilly areas where the plants are not pruned repeatedly the protein content gradually decreases from February onwards, and minimum values are found in December when leaf fall starts. Like nitrogen, phosphorus also shows the same trend, its maximum content being in February.

Mulberry leaf stalks and leftovers, after silkworm feeding, are generally fed to cattle. Their composition on a DM basis is 11.5 percent CP, 34.0 percent CF, 76.5 percent total carbohydrate, 9.3 percent total ash, 1.6 percent Ca and 0.2 percent P. The calculated values were 70.8 percent for digestible CP, 48.4 percent for total digestible nutrients and 35.6 percent for starch equivalent on a dry basis. The balance of N, Ca and P were positive and animals also gained weight.

A trial feeding of mulberry leaves to sheep was conducted by Jayal and Kehar. The digestible nutrients of leaves on a DM basis were 10.7 percent for CP, 8.3 percent for CF and 39.6 percent for total digestible nutrients. Incorporation of 6 percent shade-dried mulberry leaves with poultry feed increased egg production.

To ascertain the effect of feeding leftover mulberry leaves with stalks, a study was conducted by the authors during 1999-2000 at Kolar, the traditional sericultural area of Karnataka (India) with hybrid milk cows (local x Jersey) at the farmer level. Twenty farmers, who regularly feed mulberry leftover leaves and stalks after silkworm feeding, were selected. In general, only one feeding of mulberry leaves at noon was carried out. There was a control group of ten cows.

On average, 15-20 kg of either full mulberry leaves (if there were any in excess after rearing) or leftover leaves with stalks were fed to the cows. In the case of the control group, the same feeding schedule was followed except at noon, when mulberry leaves were replaced by cereal straw. The amount of feed supplied at different times is shown in Table 1.

Table 1. Schedule of feeding and quantity of feed given to milk cows.

Feeding time	Control feed	kg	Mulberry feed	kg
06:00h	Fodder grass	15	Fodder grass	15
11:00h	Finger millet grass	20	Finger millet grass	20
13:00h	Rice straw	15	Mulberry	15
22:00h	Concentrate	10	Concentrate	10

Farmers provided mulberry leaves (either leftovers or full) only once a day. It was found that animals did not have any reluctance in accepting mulberry as feed. Records were kept on milk yield, fat percentage and time required for next pregnancy for both the groups. The data collected were subjected to statistical analysis (T test) to study the significance of feeding mulberry leaves. The total quantity of feed was constant for both groups (Table 1). Animal performance data are presented in Table 2.

The study clearly shows that mulberry improved feed quality significantly. Nearly 34 percent higher milk yield was recorded when mulberry was provided once a day. Milk yield was found to increase both morning and evening, although the improvement was found to be more pronounced evening milking.

Table 2. Comparison of normal feeds vis-à-vis partial mulberry feeds with respect to economic characters in hybrid cows (local x Jersey)

Group	Milk yield (litres)			Fat %	Days to conception
	Morning	Evening	Total		
Control (n = 10)	8.4	6.7	15.1	2.94	100
Mulberry (n = 20)	11.0	9.2	20.2	3.50	77
"t" value	9.203	9.366	9.747	21.43	19.56
Significance	P<0.01	P<0.01	P<0.01	P<0.01	P < 0.01

Fat content was higher in cows fed mulberry (3.50 against 2.94 percent). Interestingly, cows fed mulberry conceived 24 days. However, 25 percent of farmers believe that continuous feeding with mulberry can cause termination of pregnancy in some cases for unknown reasons. Experiments are required to study this theory.

Nutritional Value of Mulberry

Dairy husbandry phases the challenge of improving production and feeding systems. It is hoped that future models have as their basis a sustainable focus and a rational use of natural resources. Mulberry, although originally from Asia, has adapted excellently to the tropics of Central America and gives very high yields. For forage production, mulberry has shown excellent characteristics of palatability and, consequently, high intakes by cattle. The whole plant contains between 14-22 percent crude protein on DM basis and *in vitro* digestibility between 70 to 80 percent.

The mulberry samples analysed by the UFAG Laboratories in Switzerland, from the high zone of Cartago in the Central Plateau of Costa Rica, gave on a

DM basis, 22 percent of crude protein (CP), 19 percent of crude fibre (CF), 2.3 percent of ether extract, 50 percent of by-pass protein and an estimated 1.48 Mcal/kg of net energy for lactation. Eswara and Reddy when evaluating the nutritive value of mulberry leaves, reported DM intakes of 2.74 and 3.55 percent of liveweight in goats and sheep, respectively, with 12 percent CP and 71 percent TDN.

The nutritional value of mulberry is in its leaves. However, being a shrub, biomass supply is considered an important aspect when used in animal feeding. Boschini, Dormond and Castro reported leaf:stem ratios above 1, before 100 days of growth. Leaf:stem proportion varies with cultivation and management conditions, plant density and cutting frequency. The whole mulberry plant, once chopped, is offered to cattle. This represents a mixture of leaf and stem which the animal selects during consumption. For this reason it is necessary to know precisely chemical composition of leaves and stems. The results of the influence of plant density, cutting height and frequency on the chemical composition throughout the year in a high tropical environment are presented here.

Bromatological Composition of Forage

The chemical composition of leaves and stem as affected by planting density, cutting height and frequency are presented in Table 3. DM percentage in the fresh forage was very constant ($P > 0.05$) in leaves for the different planting densities. Cutting height and frequency did not influence DM percentage ($P > 0.05$). However, the difference in DM content between 30 and 60 cm in cutting height was less than 1 percent. There were no differences in DM percentage between 56 and 84 days, and only 2.3-3 percent between them and 112 days.

Table 3. Leaf and stem chemical composition (% DM) of mulberry at three spacings, two cutting heights and three frequencies

Factor	*DM*	*CP*	*NDF*	*Hem.*	*ADF*	*Cel.*	*Lignin*	*Ash*
				Leaf				
Spacing								
60 cm	24.5	23.7	33.5	9.2	24.4	18.5	5.8	16.7
90 cm	24.5	23.5	32.4	8.3	24.1	19.1	5.1	24.5
120 cm	24.8	23.2	32.6	8.6	24.1	18.7	5.4	24.8
Height								
30 cm	24.1	23.8	33.5	9.4	24.1	18.9	5.3	17.0
60 cm	25.1	23.1	32.1	7.9	24.2	18.7	5.6	16.5
Frequency								
56 days	24.1	25.6	34.3	9.1	25.2	19.5	5.7	16.1
84 days	23.7	22.2	32.3	8.8	23.5	18.6	4.9	16.9
112 days	26.8	20.8	30.7	7.6	23.1	17.4	5.7	18.1
Stem								
Spacing								
60 cm	23.2	8.7	63.2	14.5	48.7	40.0	8.7	8.2
90 cm	23.4	8.8	63.2	14.5	48.8	40.5	8.3	8.3
120 cm	24.9	8.8	64.4	16.3	48.1	39.6	8.5	7.7
Height								
30 cm	23.4	8.8	63.6	14.8	48.8	40.2	8.6	8.0
60 cm	24.2	8.8	63.6	15.4	48.2	39.8	8.4	8.1
Frequency								
56 days	20.9	10.7	59.0	15.7	36.6	43.3	6.7	10.0
84 days	23.6	7.8	65.9	12.7	43.2	53.2	10.0	7.1
112 days	30.0	6.3	69.8	17.0	42.7	52.7	10.1	5.5

Crude protein content showed very little differences (P <0.05) between spacings and cutting heights. These

Table 4. Leaf and stem chemical composition (% dry matter) of mulberry cultivated at three planting distances and harvested at three cutting frequencies.

Spacing (cm)	Frequency (days)	DM	CP	NDF	Hem.	ADF	Cel.	Lignin	Ash
					Leaf				
60	56	24.4	25.9	35.0	10.1	24.8	19.3	5.5	16.3
60	84	23.2	22.5	33.5	9.6	23.9	18.4	5.4	16.6
60	112	26.3	20.9	30.6	6.5	24.1	16.8	7.2	17.6
90	56	24.8	25.8	33.4	8.6	24.9	19.3	5.5	16.1
90	84	23.2	22.0	31.3	7.3	24.1	18.6	4.4	17.0
90	112	25.6	21.0	31.6	8.9	22.6	17.7	4.9	18.0
120	56	23.1	25.1	34.5	8.6	25.9	19.9	6.0	15.9
120	84	24.6	22.3	31.9	9.5	22.5	17.8	4.8	16.9
120	112	28.5	20.6	29.8	7.3	22.5	17.6	4.9	18.6
				Stem					
60	56	21.4	10.6	58.5	15.7	42.8	36.4	6.4	9.9
60	84	21.2	7.8	64.9	11.4	53.4	42.9	10.5	7.6
60	122	29.6	6.3	70.3	16.1	54.3	43.4	10.9	5.5
90	56	21.2	10.7	58.6	14.9	43.7	37.1	6.7	10.2
90	84	23.0	7.8	66.1	12.3	53.8	44.3	9.4	7.2
90	112	28.2	6.3	68.7	16.6	52.1	42.2	9.9	5.7
120	56	20.1	10.8	59.8	16.5	43.3	36.2	7.1	9.7
120	84	26.6	7.6	66.7	14.4	52.4	42.4	10.0	6.5
120	112	32.1	6.3	70.2	18.3	51.9	42.4	9.5	5.3

Table 5. Leaf and stem chemical composition (percent of DM) of mulberry cultivated at three planting distances and harvested at two cutting heights

Spacing (cm)	Frequency (days)	DM	CP	NDF	Hem.	ADF	Cel.	Lignin	Ash
				Leaf					
60	30	24.6	33.9	33.9	9.6	24.2	18.8	5.4	17.3
60	60	24.4	23.4	33.2	8.6	24.5	18.3	6.3	16.1
90	30	23.5	23.7	32.0	8.2	23.8	18.7	5.1	17.2
90	60	25.6	23.3	32.8	8.3	24.4	19.4	5.0	16.4
120	30	24.3	23.5	34.8	10.4	24.4	19.1	5.4	16.5
120	60	25.3	22.9	30.5	6.8	23.7	18.3	5.4	17.2
				Stem					
60	30	23.0	8.8	62.7	13.7	49.0	40.3	8.7	8.3
60	60	23.5	8.7	63.7	15.2	48.4	39.8	8.6	8.0
90	30	23.3	9.0	63.2	15.2	48.0	39.6	8.4	8.2
90	60	23.5	8.7	63.3	13.8	49.5	41.3	8.1	8.3
120	30	24.0	8.7	64.9	15.5	49.4	40.7	8.6	7.5
120	60	25.8	8.9	63.8	17.1	46.8	38.4	8.4	7.9

variations were accentuated (P <0.01) between cutting frequencies, decreasing two to three percent for every 28 days from eight weeks growth. Planting densities did not influence (P >0.05) cell walls, ADF, hemicellulose, cellulose, lignin and total ash. Cell wall and total ash showed very small differences between cutting heights. Hemicellulose, ADF, cellulose and lignin showed not significant differences (P >0.05) between cutting heights. Cutting height influenced cell wall and total ash (P <0.01), ADF and cellulose (P <0.05). Changes in hemicellulose and lignin were not significant (P>0.05) with height.

Table 6. Leaf and stem chemical composition (percent of DM) of mulberry harvested at two cutting heights and three cutting frequencies

Height (Cm)	*Freq. (days)*	*DM*	*CP*	*NDF*	*Hem.*	*ADF*	*Cel.*	*Lignin*	*Ash*
				Stem					
30	56	23.6	25.8	35.0	10.0	25.0	19.7	5.4	16.5
30	84	23.0	22.8	34.0	10.2	23.8	18.5	5.4	17.0
30	112	26.8	21.0	29.9	7.2	22.8	17.7	5.2	18.1
60	56	24.6	25.4	33.6	8.3	25.4	19.3	6.0	15.7
60	84	24.4	21.7	30.5	7.4	23.1	18.7	4.4	16.7
60	112	26.9	20.6	31.4	8.0	23.4	17.2	6.2	18.0
				Stem					
30	56	21.5	10.7	58.9	15.2	43.7	37.0	6.7	9.9
30	84	22.0	7.9	66.0	12.3	53.7	43.4	10.2	7.2
30	112	29.2	6.2	69.7	17.2	52.6	42.4	10.2	5.4
60	56	20.3	10.7	59.1	16.2	42.9	36.2	6.7	10.0
60	84	25.2	7.6	65.8	13.1	52.7	43.0	9.7	7.0
60	112	30.7	6.4	69.8	16.9	52.9	42.9	10.0	5.6

The interaction planting density and cutting frequency was significant (P <0.01) for leaf DM content and its content of protein and minerals. The structural fractions were not affected (P <0.01) by the group effect (Table 4). In Table 5, the chemical composition of leaves and stems planted at different spacing and harvested at different heights are presented. DM, CP, cell walls, hemicellulose and total ash all had an interaction effect (P <0.05) due to these variables. ADF and its components were not affected. The associated effects of cutting height and frequency (Table 6) on chemical composition were significant for CP, cell wall (P <0.01) and total ash (P <0.01).

Contents of DM, NDF components and hemicellulose did not show important variations. Table 7 presents the values of leaf chemical components detailed by cutting number within each frequency. The effect was highly significant (P<0.01) for all chemical fractions studied. The highest level interaction, density by cutting height, influenced DM, CP and total ash (P <0.01). None of the structural carbohydrates showed important variations.

Table 7. Leaf and stem chemical composition (percent of DM) of mulberry harvested at three cutting frequencies by cut number

Freq. (days)	*Cut #*	*DM*	*CP*	*NDF*	*Hem.*	*ADF*	*Cel.*	*Lignin*	*Ash*
				Leaf					
56	1	22.0	27.9	37.4	15.3	22.0	17.8	4.2	15.4
56	2	22.0	27.6	46.0	10.2	35.8	28.6	7.2	16.7
56	3	26.6	23.9	33.9	9.8	24.1	17.9	6.3	16.0
56	4	25.6	24.5	27.8	5.4	22.4	16.5	5.9	16.6
56	5	24.0	24.1	30.6	7.4	23.2	17.5	5.6	16.8
56	6	24.4	25.7	30.2	6.5	23.6	18.7	4.9	15.0

Contd.....

Contd.....

84	1	15.8	24.7	34.5	10.4	24.1	18.7	5.4	15.8
84	2	25.0	21.1	35.7	12.7	23.0	18.1	5.1	17.9
84	3	26.2	22.1	26.9	4.1	22.9	18.5	4.4	17.3
84	4	27.8	21.0	31.8	8.0	23.8	19.2	4.6	16.5
112	1	26.3	22.1	32.8	8.9	23.9	16.9	7.0	17.3
112	2	26.9	20.6	28.3	6.3	22.0	16.7	5.4	19.6
112	3	27.2	19.8	30.9	7.6	23.3	18.7	4.6	17.4
				Stem					
56	1	17.3	13.4	63.2	18.1	45.2	38.3	6.9	9.7
56	2	17.1	11.3	60.1	35.1	25.1	19.4	5.7	10.1
56	3	21.1	10.3	57.6	11.5	46.1	38.5	7.6	10.3
56	4	23.6	9.8	57.9	10.5	47.4	41.0	6.4	10.5
56	5	22.4	9.44	58.5	9.3	49.2	42.5	6.7	9.7
56	6	24.0	10.0	56.7	9.8	46.9	39.8	7.1	9.4
84	1	13.8	8.9	68.9	14.5	54.3	44.2	10.1	6.9
84	2	22.3	8.6	62.7	12.0	50.7	41.5	9.2	8.2
84	3	25.7	7.0	65.6	11.0	54.6	43.7	10.9	6.8
84	4	32.6	6.5	66.4	13.3	53.2	43.5	9.7	6.6
112	1	27.3	6.9	70.4	19.3	51.1	41.9	9.2	5.2
112	2	28.2	6.9	69.2	14.9	54.3	43.7	10.6	6.0
112	3	33.9	5.2	69.6	16.8	52.8	42.4	10.4	5.2

DM in stems did not show important variations (P >0.05) between planting densities. Between cutting height there was a difference (P <0.05) of less than 1 percent and between cutting frequencies there were significant variations (P <0.01). CP content was very constant (P >0.05) between densities and cutting heights. Cutting frequency influenced (P <0.01) CP very clearly. Cell wall, ADF and its components cellulose and lignin, did not show important variations (P <0.05) between planting densities, but hemicellulose and total ash were affected slightly (P<0.05). Cutting height did not influence stem structural carbohydrates or total ash. Cutting frequency had a marked effect (P <0.01) on cell wall and its components, and on total ash.

The interaction between density and cutting frequency (Table 4) was significant (P <0.01) for stem DM but CP, ash and structural carbohydrates were not. Table 5 shows the chemical composition of the stem of plants at different densities and cutting heights. Hemicellulose, ADF and cellulose were affected by the associated action of these factors significantly (P <0.01). Other components were not affected. When analysing the interaction of cutting height and frequency on stem chemical composition (Table 6), only DM was affected (P <0.01).

Table 7 presents stem chemical composition by harvest number at each cutting frequency. The analysis showed that harvest numbers within each frequency had a significant effect (P<0.01) on all chemical components studied. The interaction between density and cutting frequency only affected stem DM content (P <0.05). The general chemical composition (mean and standard deviation) of the experiment was: DM 24.6 percent (±2.24); CP 23.5 percent (±0.59); NDF 32.8 percent (±3.87); hemicellulose 8.7 percent (±3.85); ADF 24.2 percent (±2.76); cellulose 18.8 percent (±2.46); lignin 5.4 percent (±2.21) and total ash 16.8 percent (±0.75) in leaves and DM 23.8 percent (±2.01); CP 8.8 percent (±0.66); NDF 63.6 percent (±2.8); hemicellulose 15.1 percent (±2.8); ADF 48.5 percent (±2.99); cellulose 40.0 percent (±2.78); lignin 8.5 percent (±1.13) and total ash 8.0 percent (±0.88) in the stems.

Planting density produced significant variations on leaf CP and stem DM, hemicellulose and total ash. However, differences were slight from the biological point of view. The rest of the components showed very narrow variation from the mean. In general, spacing did not influence the nutrients present in leaf and stem. Cutting height statistically also affected leaf DM, CP, cell wall, ADF and total ash, but only stem DM. These differences were less than 1 percent, which are not biologically important to favour a certain cutting height.

Contrary to what had been found for planting density and cutting height, cutting frequency had a significant effect on nutrient accumulation, except for leaf hemicellulose and lignin. At frequent cuttings, leaf and stem CP and leaf cell wall were high. At longer cutting, stem NDF and all cell wall components increased. Total ash in leaves increased with cutting interval, whereas the opposite occurred with the stem.

Chemical composition was very different between leaf and stem, except for the percentage of DM. Leaf CP was three times that of the stem, and leaf cell wall and ADF were half. Cellulose was 75 percent of ADF in leaf and 80 percent in stem. Lignin was 25 percent of ADF in the leaf and 20 percent in the stem. Mulberry had two to three times more ash in leaves than in stem. Piccioni reports, on DM basis, mulberry leaf CP of 16 percent, ether extract 4.1 percent, cellulose 6.9 percent and ash 11 percent. Rodríguez, Arias and Quiñones found CP concentrations of 21 percent without fertiliser and 24 percent with 80 kg of N/ha after each 6-week harvest.

Increasing cutting frequency to 12 weeks, leaf CP decreased to 18 percent without fertiliser and to 16 percent with fertiliser. Eswara and Reddy reported DM and CP values of 30 percent and 14.5 percent respectively. When comparing mulberry with tropical and subtropical grasses, CP is higher than in alfalfa (17 percent), young Orchard grass (15 percent), Pangola grass (1 percent) and

Elephant grass (9 percent). Mulberry leaf has 33 percent NDF and the stem 64 percent, whereas alfalfa has 40 percent, Orchard grass 55 percent, Pangola grass 70 percent and Elephant grass 72 percent, Mulberry stem at 112 days has a cell wall equal to or smaller than any tropical grass.

Lignin in subtropical grasses varies between 4-8 percent and in tropical grasses between 7 and 8 percent, whereas mulberry has 5.5 percent in the leaves and 8 percent in the stem. It can be deducted that mulberry leaf is superior to any subtropical forage and that mulberry stem compares to tropical grasses. Considering that net energy for lactation in cattle is calculated from TDN from lignin content in NDF or ADF, the high digestibility of mulberry is confirmed. Lignin represents between 15 and 18 percent of leaf cell wall and between 11 and 15 percent of stem cell wall, which is used to estimate the high digestibility of both fractions.

The main limitation of the use of mulberry stem as feed is not related to chemical quality per se but to the physical form in which it is offered to animals for their use. Many tropical feeds, rspecially in dry periods, are inferior to mulberry stem in terms of quality. Mulberry leaf, properly dried and ground, is of excellent quality and its characteristics make it most suitable for inclusion in compounded feed for highly producing dairy cattle.

Cultivation

The ideal soil for mulberry should have medium texture, with 60 cm in depth, phreatic level of at least 1 m, good drainage, pH 6-7 and good organic matter. Mulberry does not tolerate flooding. Mulberry is reproduced asexually. Stakes should be 15-20 cm in length and 18-20mm in diameter. It is important to use matured stems (more than three months of age), with stakes chosen from the middle part.

Direct planting is not recommended because of high failure, high weeding cost and low resistance to the dry period. Rooting is done on 1.20 m x 0.10 m beds, where stakes are planted with a spacing of 10 x 20 cm, burying two-thirds of the stake. Beds are covered with a plastic sheet to control humidity and weeds. After 45 days, the plastic is cut to allow aeration. The operation should last two to three months. Transplanting is carried out by pulling the plants from the bed, removing the leaves and once planted, by pruning to 10-15 cm.

Recommended density lies between 25 and 32 000/ha (e.g. 80 x 40 cm; 80 x 50 cm; 90 x 40 cm). It is advisable to apply 50 kg of calcium phosphate and 400 g of organic matter per plant at transplanting. Strict weed control is required during establishment since weeds are one of the mayor problems of the forage crop.

Three prunings are recommended: formation pruning (to 20 cm) after two or three months of transplanting, to shape the plant (productive crown) and to encourage shooting; regular harvests at 20-40 cm, 1 cm higher than the previous cut, every 75-90 days; and rejuvenating pruning (to 10-20 cm) after the plant has formed a multiple-shoot high crown called "deer's antlers" which has low productivity.

Forage production increases up to the third year, when yield reaches 30-50 tonnes of fresh leaves per year. At Hacienda Lucerna, the average production is 50t/ha/year of fresh leaves with 90 days cuts. In 1993, an evaluation was initiated with the combination of mulberry with Gliricidia sepium, mixed planted at a spacing of 40-60 cm between plants in double lines 1 m apart. The objective was to profit from the N fixation by

Gliricidia. With this system annual yields of fresh foliage of 60t/ha of mulberry and 30t/ha of Gliricidia, have been obtained.

Nutritive Value

Digestibilities between 80 and 93 percent have been reported for mulberry leaves and 50 percent for stems. Velázquez found 205 ppm of Ca, 55 ppm of P, 6.0 ppm of Fe and 0.4 ppm of Zn. Mulberry leaves (Kanva 2) from Hacienda Lucerna had 20.0 percent DM, 15.8 percent CP, 11.5 percent CF, 4.6 percent EE, 50.9 percent N-free extract and 17.3 percent ash, with a calculated net energy value of 2.4 Mcal/kg.

11

Mulberry in Animal Production

Mulberry is a shrub or a tree traditionally used for feeding the silkworm in various countries. It belongs to the Order Urticales, Moraceae Family and genus *Morus*. The better known species, *Morus alba L.* and *Morusnigra L.*, seem to have their origin in the Himalayas' foothills. It has excellent nutritional value as forage.

In Spain, Mulberry was recommended to be planted in association with other crops like maize, potatoes, vegetables, alfalfa and fruit trees, always controlling spacing and pruning to avoid light competition. Some authors recommend planting at 80 cm between plants and rows. In other articles, densities of 30,000 plants per ha for low pruning (below 70cm); 7-12,000 plants for medium pruning (70-170cm); and between 2,250 and 6,000 plants for high pruning (above 170cm) are mentioned.

The available yield information is almost exclusively on leaves, since it is the part used to feed the silkworm. In France, fresh leave yields of 17,000 kg/ha are reported with 7 x 7 m spacing. With higher densities, yields of 30,000 kg/ha have been obtained. Yields are related to plantation age and more specifically with trunk diameter. These authors report that annual leaf production in monoculture increases from 6,500 kg to 33,500 kg/ha from the first to the seventh year.

In good land, leaf yield per plant varies from 9 to 70 kg when average trunk diameter increases from 7 to 55 cm. With 22.5 tons of human faeces and 300kg of ammonium sulfate, fresh leaf production can reach 13 ton/ha/year. In Paraguay, leaf yields of 20,000 kg/ha have been obtained in 4-year plantations harvesting at 30cm from the surface.

Work with widely spaced plants in Turrialba (Costa Rica) a yield of 2.32 kg of dry matter per plant was calculated per year cutting at a height of 50 cm. Cutting at 1m, the yield decreased to 2.16kg. However, leaf production was 1kg for both cutting heights. With cuts every 60,120 and 180 days, total dry matter production was 1,64, 2,17 and 2,86 kg/plant/year, respectively. However, leaf production declined from 1.11 to 0.84kg between 60 and 180 days.

Nutritional Value

The nutritive value of Mulberry is one of the highest found in products of vegetable origin, far superior to traditional forages like alfalfa. Mulberry biomass is remarkable due to one characteristic which is found in very few plants; high levels of CP and high levels of digestible energy.

The content of dry matter (DM) and other components in the leaves of Mulberry is higher when compared to traditional grasses used in animal feeding (Table 1). From Costa Rica there are reports of 25-32% DM in leaves; 23-29% for young stems; and 24-45% for woody stems.

Table 1. Dry matter, crude protein and IVDMD of Mulberry foliage and other feeds used in Central America

Item	DM,%	CP,%	IVDMD,%
Mulberry (*M. alba*)	28.7	23.0	80.0
King grass (*P. purpureum x P. typhoides*)	20.0	8.2	52.7
Star grass (*Cynodonlnem-fluensis*)	22.3	8.9	54.9
Commercial concentrate	91.5	17.7	85.0

In plantations of three Mulberry varieties planted at 0.40m between plants and 1m between rows (25,000 plants/ha), in three ecologically different sites in Costa Rica and with various fertilisation levels, it was found that DM content of leaves and edible stem were more affected by the location than by fertilisation level, without differences among varieties (Table 2).

Table 2. Site and nitrogen fertilisation effects on leaf dry matter content in three Mulberry (M. alba) varieties in Costa Rica

Site	Variety, % DM			
	Criolla	Indonesia	Tigriada	Mean[2]
Puriscal	31.6	32.0	30.7	31.4[b]
Coronado	26.8	25.7	22.2	24.9[c]
Paquera	29.0	36.0	31.9	32.3[a]
N[1], kg/ha/year				
180	29.2	31.4	28.8	29.8[a]
360	28.9	31.5	28.6	29.7a[b]
540	29.2	30.8	27.4	29.1[b]
Mean[2]	29.1[b]	31.2[a]	28.3[c]	

1 = Ammonium nitrate

2 = Values with same letter are not statistically different $p < 0.05$

In the woody stems, the differences were still more pronounced, with values between 27 and 48% of DM (Table 3). The small effect of fertilisation on DM was also reported in a plantation where various levels of goat manure were applied (Table 4).

Table 3. Site and nitrogen fertilisation effects on woody stem dry matter content in three Mulberry (M. alba) varieties in Costa Rica

Site	Variety, % DM			
	Criolla	Indonesia	Tigriada	Mean[1]
Puriscal	45.2	41.9	42.9	43.4[b]
Coronado	31.6	27.1	24.2	27.6[c]
Paquera	48.8	47.8	48.8	48.5[a]
N, kg/ha/year				
180	42.7	39.2	39.8	40.6[a]
360	41.8	39.9	38.4	40.0[a]
540	41.2	37.8	37.8	38.9[b]
Mean1	41.9[a]	38.9[b]	38.6[b]	

1 = Values with same letter are not statistically different, $p<0,05$.

Table 4. Dry matter content of Mulberry (Morus alba) by soil manure application

Fraction %	Manure level[1]				NH_4NO_3[1]
de MS	**0**	**240**	**360**	**480**	**480**
Leaves[2]	26.4[a]	25.9[bc]	26.0[b]	25.5[d]	25.6[cb]
Young stem	28.0[a]	27.1[b]	26.8[b]	26.0[q]	24.7[d]
Woody stem	42.2[a]	41.5[b]	40.8[c]	40.4[cd]	40.0[d]

1 = kg de N/ha/year.
2 = Values with same letter horizontally do not differ statistically, $p<0,001$.

Mulberry leaves have high levels of CP and IVOMD when compared to other feed used in ruminant feeding. Data from Central America indicate CP values between 15-25% and IVOMD between 75-90%, which indicates a quality comparable or superior to commercial concentrates (Table 5). Non lignified stem also has a good nutritional quality, with CP values of 7-14% and IVOMD of 56-70%.

Table 5. Site and nitrogen fertilisation effect on crude protein in three Mulberry (Morus alba) varieties in Costa Rica.

Site	Variety, % PC			
	Criolla	Indonesia	Tigriada	Mean
Puriscal	20.1	22.2	21.0	21.1[b]
Coronado	22.2	26.3	25.7	24.8[a]
Paquera	17.0	14.8	13.7	15.1[c]
N kg/ha/year				
180	18.9	20.8	20.0	19.9[a]
360	20.5	20.8	19.9	20.4[a]
540	20.0	21.6	20.6	20.7[a]
Average	19.8[a]	21.1[a]	20.1[a]	

Table 6. Site and nitrogen fertilisation effect on crude protein of young stem in three Mulberry (Morus alba) varieties in Costa Rica

Site	Variety, % CP			
	Criolla	Indonesia	Tigriada	Mean
Puriscal	11.2	13.0	12.1	12.1[a]
Coronado	13.4	13.9	14.3	13.9a
Paquera	9.4	10.2	8.7	9.4[b]
N, kg/ha/year				
180	11.2	12.7	11.6	11.4[a]
360	11.8	11.8	11.8	11.7[a]
540	11.3[b]	12.6[a]	11.7[b]	11.9[a]
Mean	11.3[b]	12.4[a]	11.7[b]	

As it was the case for DM, CP and IVDMD is similar among varieties and is not affected much by fertilisation nor by cutting frequency (Tables 6, 7, 8, 9, 10 and 11). Although, an effect was observed on CP content of leaves and young stem when ammonium nitrate was added instead of goat manure at iso-nitrogenous levels (Table 9).

Table 7. Site and nitrogen fertilisation effect of leaf IVDMD in three Mulberry (Morus alba) varieties in Costa Rica

Site	Variety, % IVDMD			
	Criolla	Indonesia	Tigriada	Mean
Puriscal	78.0	75.5	76.8	76.7[a]
Coronado	75.4	73.8	75.6	74.9[a]
Paquera	71.6	71.2	71.6	71.5[b]
N, kg/ha/year				
180	74.9	72.6	75.0	74.2[a]
360	74.8	73.6	74.1	74.2[a]
540	75.3	74.2	74.9	74.8[a]
Year	75.0[a]	73.5[a]	74.7[a]	

Table 8. Site and nitrogen fertilisation effect on IVDMD of young stem in three Mulberry (Morus alba) varieties in Costa Rica.

Site	Variety, % IVDMD			
	Criolla	Indonesia	Tigriada	Mean
Puriscal	69.5	70.5	63.9	68.0[a]
Coronado	70.3	68.6	69.9	69.6[a]
Paquera	65.9	64.7	59.4	63.3[b]
N, kg/ha/year				
180	68.5	68.2	65.4	67.4[a]
360	68.9	69.0	64.2	67.4[a]
540	68.2	66.6	63.7	66.2[a]
Average	68.6[a]	67.9[a]	64.4[b]	

Table 9. Manure application effect on crude protein content of Mulberry leaf and young stem

Fraction	Manure level[1]				NH_4NO_3[1]
	0	**240**	**360**	**480**	**480**
Leaves[2]	19.1[c]	19.3[b]	19.3[c]	20.2[b]	22.5[a]
Young stem	7.4[b]	7.1[b]	7.2[b]	7.4[b]	12.3[a]

1 = Equivalence in kg of N/ha/year.
2 = Values with same letter horizontally do not differ statistically, p<0,01.

Table 10. Manure application effect on dry matter digestibility of Mulberry leaf and young stem

Fraction	Manure level[1]				NH_4NO_3[1]
	0	**240**	**360**	**480**	**480**
Leaves	76.7	77.5	77.0	76.9	77.1
Young stem2	56.2[b]	56.4[b]	56.0[b]	55.8[b]	58.6[a]

1 = Equivalence in kg of N/ha/year.
2 = Values with same letter horizontally do not differ statistically, p<0,01.

Table 11. Annual cutting frequency effect on digestibility and crude protein content of Mulberry biomass

Fraction	% CP		% IVDMD	
	3	4	3	4
Leaves[1]	19.2[b]	20.9[a]	77.2	76.9
Young stem	8.1	8.5	56.9	56.2

1 = Values with same letter horizontally do not differ statistically, p<0,01.

In Paquera, located in the Pacific Coast of Costa Rica, with high luminosity and high temperatures, leaf CP and IVDMD (15.1 and 71.5%, respectively) are reduced compared to higher locations with more clouds and lower temperatures ((24,8 y 74,9%, respectively), like is the case of Coronado and Puriscal, located in the mountainous areas of the country. The greater luminosity and higher temperature in Paquera can explain the lower water content in all fractions, lower CP and IVDMD levels, and high Dm production, as will be seen later.

It is known that high luminosity reduces nitrate levels and increases cell wall components and growth due to larger photosynthetic activity. Coronado has lower temperatures and more cloudiness and rainfall, which could have limited growth and lignification. But the high soil fertility and the lower growth rates of this site, explain the higher CP and lower biomass DM. In addition to the climatic factors of Paquera, the lower N, copper and zinc soil contents could limit fertility and nutrient content in the plant. On the other side, the clay soils, the lower pH and the low potassium contents can explain low biomass yields in Puriscal.

Working with goats fed exclusively on Mulberry and "Amapola" leaves confined to metabolic cages, apart from high DM intake values, high levels of *in vivo* DM and CP digestibilities of Mulberry leaves (Table 12). The values reached 90% for leaf CP digestibility.

Table 12. Dry matter and crude protein in vivo digestibilities of Mulberry and "Amapola" (Malvaviscus arboreus) with goats in metabolic cages

Forage species	Parameters		
	Intake, % LW	IVDMD, %	IVCPD, %
Mulberry	4.18	79.3	89.5
Amapola	4.35	64.2	54.2

Although statistical comparisons have not been made, nitrogen, potassium and calcium contents of leaves and young stems are high (Tables 13 and 14), reaching values of 3.35, 2.0 and 2.5%, respectively. In other work (Table 15), no appreciable differences were found in mineral content of leaves and young stems when increasing amounts of legume leaves were added to the soil .

In a report of the proximate analysis of Kanvas-2 variety of Mulberry (Table 16), levels of protein (29.6%), ash (7.53%) and nitrogen-free extract (50.0%) were considered high. At the same time, low levels of crude fibre were reported (10.1%). In this work, the type and quantity of leaf amino acids were determined. Twenty four amino acids in significant concentrations were detected, plus 6 more at low concentrations. The low fibre level and high contents of CP and IVDMD justify future evaluations of Mulberry as an ingredient for high quality meal and compounded feeds.

Table 13. Fertilisation effect on the mineral content of Mulberry (Morus alba) leaf in Costa Rica

Variety	Minerals, % of Dry Matter				
	N	P	K	Ca	Mg
Criolla	3.17	0.30	2.07	1.90	0.47
Indonesia	3.37	0.30	1.73	2.87	0.63
Tigriada	3.23	0.40	2.33	2.74	0.55
N, kg/ha/year					
180	3.18	0.33	1.96	2.54	0.55
360	3.26	0.33	2.05	2.49	0.55
540	3.32	0.33	2.12	2.48	0.55
Mean	3.26	0.33	2.04	2.50	0.55

Table 14. Fertilisation effect on mineral content of Mulberry (Morus alba) young stem in Costa Rica

Variety	Minerals, % of Dry Matter				
	N	P	K	Ca	Mg
Criolla	1.81	0.30	2.24	1.33	0.39
Indonesia	1.98	0.40	2.99	1.53	0.49
Tigriada	1.87	0.33	2.41	1.38	0.40
N, kg/ha/year					
180	1.89	0.33	2.48	1.43	0.44
360	1.86	0.33	2.60	1.40	0.43
540	1.90	0.37	2.55	1.42	0.41
Mean	1.89	0.34	2.55	1.41	0.43

Table 15. Mulberry leaf mineral content as affected by level of Poro foliage added to the soil

Mineral	Proportion of Poro foliage in the soil				
	Control Without trees				Mean
		0%	50%	100%	
N	2.90	3.16	3.10	3.09	3.06
P	0.22	0.22	0.22	0.23	0.22
K	1.51	1.37	1.35	1.71	1.49
Ca	1.64	1.82	1.74	1.77	1.74
Mg	0.42	0.44	0.41	0.38	0.41

Table 16. Leaf composition of Mulberry variety Kanvas-1, in El Salvador, Central America

Leaf position	Fractions, % of Dry Matter					
	Dry matter	Crude protein	Crude fat	Crude fibre	Ash	NFE
Terminal	19.5	33.6	3.2	9.4	7.7	46.1
Intermediate	22.5	28.3	2.8	10.2	7.3	51.4
Basal	23.0	26.7	3.4	10.8'	7.7	52.6
Average	21.7	29.6	3.1	10.1	7.5	50.0

Animal Response

Evaluations carried out with ruminants (cattle, goats and sheep) show high intakes levels of DM and high animal responses in weight gain and milk yield. Oviedo when comparing Mulberry foliage with concentrate (Table 17) as supplements to Jersey x Criollo grazing cows, found similar milk yields with both supplement (13.2 and 13.6 kg/animal/day, respectively) at equal DM intake levels (1.0% of LW) and superior to grazing only (11.3 kg/animal/day).

Table 17. Milk yield and dry matter intake of Jersey x Criollo cows grazing Star grass (Cynodon nlemfuensis) and supplemented with Mulberry or concentrate

Feed consumed, % LW	Supplement		
	Concentrate	Mulberry	Nothing
Milk, kg/animal/day	12.45	12.08	10.34
Concentrate	1.0		
Mulberry		1.0	
Star grass	3.7	3.4	3.7
Total	4.7	4.4	3.7

Mulberry inclusion did not affect fat, protein and total solids contents in milk (Table 18), but improved net benefits when compared to the concentrate (US$ 3.29 vs. 2.84, respectively).

Table 18. Milk chemical composition from grazing cows supplemented with Mulberry or concentrates

Component	Supplement		
	Concentrate	Mulberry	Nothing
Fat	3.95	3.81	3.63
Protein	3.53	3.51	3.31
Total solids	12.53	12.39	12.31

Esquivel *et al.*, when replacing 0, 40 and 75% of the concentrate by Mulberry foliage, did not find significant differences ($p<0.05$) in milk production (14.2; 13.2 and 13.8 kg/animal/day, respectively) in Holstein cows grazing Kikuyo grass (*Pennisetum clandestinum*) without appreciable effects on milk quality (Table 19). Also in this work, considering only feeding costs, net income per animal was 11.5% higher with the maximum level of Mulberry than that obtained with concentrate only.

Table 19. Effect of the substitution of concentrate by Mulberry foliage on milk yield of Holstein cows grazing Kikuyo grass (Pennisetum clandestinum)

Parameter	Relation Concentrate:Mulberry		
	100/0	60/40	25/75
Milk, kg/animal/day	14.2	13.2	13.8
Intake, kg DM/animal/day			
Concentrate	6.4	4.2	1.9
Mulberry	0	2.8	5.5
Grass	9.3	7.8	6.2
Total	15.7	14.8	13.6

With cattle, attractive liveweight gains have been obtained when using Mulberry foliage as supplement. In the humid tropics of Turrialba (Costa Rica), Jersey x Criollo heifers grazing Star grass (*Cynodon nlemfuensis*) and supplemented with concentrate, Mulberry and concentrate or only Mulberry, no statistically differences were detected (p<0,05) among supplements (Table 20). The combination with Mulberry and concentrate gave the highest gains (742 g/animal/day).

Table 20. Intakes and liveweight gains of grazing dairy heifers supplemented with concentrate and Mulberry foliage

Parameter	Only concentrate	Concentrate + Mulberry	Mulberry
Liveweight gain	0.620	0.742	0.600
Intake of Mulberry, %LW	-	0.5	1.0
Intake of concentrate, % LW	1.0	0.5	-

With young Romo-sinuano bulls in total confinement and fed a basal diet of Elephant grass (*Pennisetum purpureum*), gains of 40, 690, 940 y 950 g/animal/day were observed with whole Mulberry DM intakes of 0, 0.90, 1.71 and 2.11% of LW as supplement. In this study, the benefit/cost relations were 0.10, 1.11, 1.18 and 0.97 for each of gain levels, respectively. The study lasted 70 days and animals were between 13-16 months old, with initial liveweight between 118 and 250 kg (Table 21).

Table 21. Effect of Mulberry supplementation on intake and liveweight gains of Romo-sinuano cattle in confinement and fed a basal diet of Elephant grass

DM intake, % LW1	Mulberry dry matter offered, % LW			
	0	1.0	1.9	2.8
Fresh Mulberry	0.00[d]	0.90[c]	1.71[b]	2.11[a]
Elephant grass	2.04[a]	1.79[a]	1.29[b]	0.95[b]
Total diet	2.04[c]	2.69[b]	3.00[a]	3.06[a]
Weight gain, g/ animal/day	0,04[c]	0,69[b]	0,94[a]	0,95[a]

[1]. *Values with same letter horizontally do not differ statistically (p<0.05) with Tukey test.*

With crossbred dairy goats of 40kg liveweight, Rojas and Benavides found milk yield increases of 2.0 and 2.5 kg/

animal/day when Mulberry supplementation was raised from 1.0 to 2.6% of LW on DM basis (Table 22). There were slight increases of fat, protein and total solids contents (Table 23). In this study, high DM intakes and additive effect of Mulberry supplementation were observed. King grass was clearly substitute by Mulberry (Table 24).

Table 22. Effect of Mulberry supplementation on milk yield (kg/animal/day) of dairy goats

Production level	Level of Mulberry DM, % LW				Mean[1]
	1.0	1.8	2.6	3.4	
High	2.04	2.38	2.51	2.47	2.35
Low	1.64	1.82	1.91	2.12	1.87
Mean[2]	1.84[b]	2.10[b]	2.21[ab]	2.29[a]	

1 = Group averages differ statistically, p<0,0001.
2 = Values with same letter horizontally do not differ statistically, p<0,0001

Table 23. Effect of Mulberry supplementation on fat, protein and total solid contents of goats fed King grass

Fraction	Intake of Mulberry DM, % LW				
	1.0	1.8	2.6	3.4	Mean[1]
Fat	3.1	3.2	3.2	3.3	3.2
Protein	3.3	3.4	3.4	3.5	3.4
Total solids	10.7[b]	11.0[b]	11.4[a]	11.2[ab]	11.1

1 = Group means differ statistically, p<0,007.

Table 24. Effect of Mulberry supplementation on dry matter intake of confined dairy goats

Forage	Dry matter intake, % LW			
Mulberry	1.0	1.8	2.6	3.4
King grass	3,2[a]	2.9[ab]	2.6[b]	2.1[c]
Total	4.2[c]	4.7[b]	5.2[ab]	5.5[a]

1 = Group means differ statistically, p<0,0003.
2 = Values with the same letter horizontally do not differ statistically, p<0,0006

With Black Belly lambs receiving a basal diet of King grass, liveweight gains of 60, 75, 85 and 101 g/animal/day were reported when Mulberry was given as supplement at 0, 0.5, 1.0 and 1.5% of LW on DM basis, respectively. In this study, rather than a substitution effect, there was an additive effect of Mulberry on total DM intake (Table 25).

Table 25. Performance of Black Belly lambs fed various Mulberry levels

Parameters	Mulberry level, DM as %of LW				Significance
	0	0.5	1.0	1.5	
Initial weight, kg	15.7	15.8	15.8	15.1	
Final weight, kg	21.5	22.6	24.4	25.6	
Gain, g/day	60c	75b	85b	101[a]	**
DM intake (kg/day)					
King grass	0.66	0.60	0.62	0.60	
Mulberry leaf	0	0.09	0.19	0.28	
Total DM intake	0.66	0.69	0.81	0.88	**
DM intake, % LW	3.54	3.72	3.99	4.34	**
DM intake, g/kg/ PV0.75	73.5	75.1	84.5	92.2	**

Table 26. Milk yield (kg/animal/day) in dairy goats fed exclusively grass and Mulberry foliage in the agroforestry model

Month	1	2	3	4	5	6	7	8	9
Goat 1	3.22	3.46	3.47	3.41	2.65	2.69	2.23	2.44	2.53
Goat 2	3.41	3.93	3.53	3.44	2.91	2.67	2.68	1.86	1.71

In a three-year evaluation (Table 26), in an agroforestry model with goats fed exclusively (at 3% LW in DM basis) with King grass and Mulberry, a lactation yield of 900 kg per 300 days was reported. This is equivalent to a mean of 3 kg/day and to 4.1 kg/day at the beginning of lactation. Forage came from a Mulberry and grass plantation associated with Poro (*Erythrina poeppigina*) measuring 1,100 m^2, fertilised with goat manure, Poro foliage and feed rejects.

During the third year, the module reached 5.0 kg/day, equivalent to 16,500 kg/ha/year (Table 27). The economic analysis indicated a benefit/cost relation of 1.27, 1.39 and 1.45 for each year, respectively (Table 28).

Ensiled Mulberry

One of the most serious problems of livestock husbandry in the tropics is the rapid decline of grass quality during the dry season. Among the most used alternatives is the conservation of forage by ensiling during the rainy season in order to use it in the dry season. However, silage is traditionally made with tropical grasses rich in fibre and low in soluble carbohydrates, which affects fermentation and results in a low quality product.

Table 27. Zootechnical parameters of confined goats in the agroforestry model in Turrialba, Costa Rica

Parameter	Average	Year1	Year2	Year3
Lactation, months	14.6[1]			
Kids per parturition	2.0			
Kid weight at birth, kg	3.7			
Milk,, kg/animal/day	2.02			
Dry period, days	37.3			
Goat weight, kg	50.0			
Total milk, kg/year		1223,0[3]	1505,9[4]	1110,5[5]
Milk production, kg				
Per module/day	4.0	3.4	4.2	5.0
Maximum monthly		42.8	71.4	122.7
Minimum monthly		145.1	189.1	207.9
Minimum per module/day		1.4	2.3	4.0
Maximum per module/day		5.2	6.1	6.7

1 = 3 lactations per goat.
2 = Data of 31.5 months.
3 = from 03/91 to 03/92.
4 = From 03/92 to 03/93.
5 = From 03/93 to 10/93.

Due to its low fibre and high level of carbohydrates, Mulberry foliage can be ensiled without additives, showing a lactic fermentation pattern and low CP losses (between 16-21% of CP in the final product) while maintaining between 66 ad 71% IVDMD. These parameters are far superior to silages made with tropical forages.

Table 28. Financial analysis (in US$) of the goat agroforestry module at Turrialba, Costa Rica

Description	Years		
	1991/92	1992/93	1993*
A. Costs			
A.1 Investments			
Mulberry x Poro plantation	4.61	4.61	2.88
Grass x Poro plantation	1.66	1.66	1.04
Facilities	16.07	16.07	9.37
Animals	50.00	50.00	31.25
Subtotal	72.34	72.34	45.21
A.2 Fixed, land opportunity cost	21.17	21.17	13.23
A.3 Variables (labour)			
Cut&carry, weeding, pruning	182.65	176.19	109.77
Forage chopping and feeding	138.45	133.55	83.20
Milking	89.05	85.90	53.52
Cleaning facilities	54.60	52.67	32.81
Manure application	26.00	25.08	15.63
Mineral salts	30.66	30.66	19.16
Antelmintic	1.40	1.40	1.40
Maintenance	6.50	6.27	3.90
Subtotal	455.31	511.72	319.39
Total cost	527.65	584.06	377.83
Updated cost	610.82	643.92	396.72
B. Income			
B.1 Milk production	672.66	813.99	549.03
Updated income	778.68	897.42	576.48
C. B - A updated	167.86	253.50	179.76
B/C	1.27	1.39	1.45

* 7.5 months of 1993

Vallejo, using 40kg plastic bags and three 30-day storage periods, compared the silage made from three tree foliages (Mulberry, Amapola and Jocote). Mulberry showed the highest levels of IVDMD, good CP content, acceptable loses of ammonium nitrogen and higher lactic acid levels (Table 29). Acetic and butyric acids levels were also high with Mulberry in the first period, but they rapidly declined (Table 30).

Table 29. Chemical characterisation of silage made from shrubs and trees

Species	% DM			% IVDMD			% CP		
Period	1	2	3	1	2	3	1	2	3
Mulberry	22.1	28.7	34.3	60.5	61.6	76.9	17.9	17.0	16.7
Amapola	28.9	30.9	35.7	51.1	55.5	68.5	16.3	16.0	16.9
Jocote	20.6	22.0	19.8	49.3	58.0	74.2	17.0	17.5	15.9

Table 30. Fermentation indicators of ensiled foliage of three woody forages

Variable	Period	Species		
		Mulberry	Amapola	Jocote
pH	1	5.4	5.0	2.9
	2	4.6	5.0	3.1
	3	5.1	5.0	2.8
% Total NH_3-nitrogen	1	2.27	0.96	1.00
	2	1.56	0.81	1.28
	3	1.26	0.16	0.91
% Acetic acid	1	5.71	1.49	1.11
	2	1.97	1.25	1.57
	3	0.85	0.16	1.10
% Butyric acid	1	6.84	0.24	0.03
	2	0.56	0.24	0.01
	3	0.32	0.02	0.00
% Lactic acid	1	5.14	4.64	0.71
	2	9.00	4.26	1.84
	3	6.22	1.38	0.63

González *et al.*, in two 9 ton Mulberry silage pits for feeding young bulls, found better fermentation indicators than Vallejo. They noted high levels of lactic acid and low concentrations of propionic and butyric acids (Table 31). pH remained close to the values reported before by Vallejo and, when opening the silo, the organoleptic characteristics were excellent: green colour and lactic smell. Nevertheless, once in the feeder the silage lost within two hours its characteristic colour and smell.

Table 31. Fermentation indicators of Mulberry silage used as supplement for Romo-sinuano cattle

Indicator	Silo 1	Silo 2	Average
pH	4.45	5.25	4.85
NH3, % of total N	7.70	4.16	5.93
Acetic acid, %	3.95	1.82	2.89
Propionic acid, %	0.27	0.08	0.18
Butyric acid, %	0.02	0.03	0.03
Lactic acid, %	12.93	17.15	15.04

The use of this Mulberry silage to supplement young bulls fed a basal diet of Elephant grass, liveweight gains of 600g/day were obtained (Table 32) with silage intake of 1.1% of LW on DM basis. However, although Mulberry silage intake was similar to that observed with other grass silages, there was a lot of refusal. The low intake was not expected due to CP and IVDMD contents. The explanation could be in the physical and chemical aspects of the silage, like high pH and high compaction which might have reduced consumption. Thos forage without additive does not reach the desirable pH levels to assure stability. Gonzalez observed this in a study with micro-silos, where after 60 days the pH never dropped below 4.5.

The conclusions that can be reached from the use of silage of woody forages relate to the species used. This observation also comes from a study with dairy goats fed silage made from three woody forages, where a very high Mulberry silage intake (near 5% of LW on DM basis) was found and a milk yield of 1.9 kg/day.

Table 32. Effect of Mulberry silage supplementation on intake and weight gain of Romo-sinuano cattle fad a basal diet of Elephant grass

Dry matter intake	Mulberry dry matter offered, % LW			
	0	0.8	1.7	2.5
Ensile Mulberry, % LW[1]	0[c]	0.66[b]	1.05[a]	1.11[a]
Total dry matter, % LW	2.16[b]	2.42[a]	2.61[a]	2.64[a]
Weight gain, g/ animal/day	117[b]	404[a]	490[a]	601[a]

1 = Values with same letter horizontally do not differ statistically ($p<0,05$), using Tukey test.

Agronomy

Mulberry also shows high biomass production, as shown in several experiments where the effects on yield and quality of cutting frequency, level and type of fertilisation and location have been study. Due to its high production capacity, Mulberry extracts a lot of nutrients from the soil. This is why, a combination of organic and inorganic fertilisers has been emphasised.

One of the obvious worries on the agronomy of Mulberry is related stake planting. For this purpose an experiment was conducted to compare three woody forages and different planting methods. In all cases Mulberry showed a high germination percentage, superior to other forages. Placing stakes in vertical position was better than horizontally.

In an evaluation of cutting height (0.5 and 1.0 m) of isolated 2 year-old plants (Table 33), differences in DM and CP production were no found, although there was a significant effect on the proportion of leaf. In this same study (Table 34), greater yields were observed with longer cutting frequencies (60, 120 y 180 days). However, this increase was due to more stem production, with leaf production no being affected.

Table 33. Dry matter yields (kg/plant/year) of Mulberry at two cutting heights

Plant fraction	Cutting height		
	0.5 m	1.0 m	p <
Total	2.32	2.16	NS
Leaf	1.00	1.02	NS
Stem	1.32	1.14	NS
Leaf fraction	0.21	0.21	NS
Leaf:stem	1.04	1.29	**

Table 34. Dry matter yields (kg/plant) of Mulberry under different cutting frequencies

Fraction	Cutting frequency, days			
	60	120	180	p <
Total	1.64^{c}	2.17^{b}	2.86^{a}	**
Leaf	1.11^{a}	1.04^{a}	0.84^{b}	**
Stem	0.52^{c}	1.08^{b}	2.01^{a}	**
Leaf fraction	0.26^{a}	0.23^{a}	0.15^{b}	**
Leaf:stem	2.11^{a}	1.06^{b}	0.45^{c}	**

Under humid tropical condition in Costa Rica, for three years a Mulberry plantation with 25,000 plants/ha, has produced 35 ton DM/ year using goat manure as fertiliser. It was also observed 20% higher production with goat manure than with ammonium nitrate, at equal levels of nitrogen application (Table 35). Significant increments in production were observed with greater manure applications. Overall yields increased 10% a year, reaching 38 ton of DM/ha by the third year. Greater total biomass yields were obtained (Table 36) with a 4 month cutting frequency compared with 3 month, although leaf yield was not different.

Table 35. Yearly total Mulberry biomass (ton of DM/ha) when applying goat manure as fertiliser and cutting every 4 months.

Year	Level of manure[1]				NH_4-NO_3
	0	240	360	480	480
12	23,0[c]	24,4[bc]	26,6[b]	31,1[a]	26,7[b]
2	21,3[c]	25,2[b]	27,6a[b]	33,4[a]	29,7[b]
3	22,9[d]	28,2[c]	32,6[b]	38,2[a]	29,2[b]

1 = kg of N/ha/year.
2 = Values with same letter horizontally do not differ statistically, p<0,001.

Table 36. Dry matter yields of Mulberry (ton/ha/year) per year and cutting frequency

Fraction	Year			Cuttings / year	
	1	2	3	3	4
Leaves[1]	10.8[ab]	10.9[a]	10.4[b]	11.0	10.4
Young stem	1.6[a]	0.9[b]	0.8[c]	1.0[b]	1.3[a]
Woody stem	11.8[c]	13.2[b]	14.4[a]	16.0[a]	10.3[b]
Total	24.2[b]	25.0[ab]	25.6[a]	28.0[a]	21.9[b]
Edible	12.4[a]	11.8[b]	11.2[c]	12.0	11.6

1 = Values with same letter horizontally do not differ statistically, p<0,001.

In another study, always under hum2id tropical conditions, Oviedo found Mulberry biomass yields of 8.0, 9.4 and 10.6 ton of DM/ha when using Poro foliage as mulch equivalent to 0, 160 and 300 kg of N/ha/year, respectively.

Working with three Mulberry varieties (Criolla, Indonesia and Tigriada) in three sites of Costa Rica (Puriscal, Coronado and Paquera), Espinoza *et al.* reported important differences among varieties in total DM yields, with one variety with almost half of the yields of the other two. Nitrogen application also affected yields according to site (Table 37).

Table 37. Site and nitrogen fertiliser effect on dry matter production of three Mulberry varieties in Costa Rica

Parameter	Variety, ton DM/ha/year			
	Criolla	Tigriada	Indonesia	Average
Site:				
Puriscal[1]	11.1	15.6	19.0	15.2[b]
Coronado	8.9	19.5	18.0	15.5[b]
Paquera	22.4	31.9	39.2	31.2[a]
Fertilization[2]				
180 kg N/ha	11.2	18.0	19.2	16.1[b]
360 "	13.7	22.8	28.3	21.6[a]
540 "	17.4	26.3	28.7	24.1[a]
Average	14.1[c]	22.3[b]	25.4[a]	

1 = Values with same letter horizontally do not differ statistically (p<0,05) with Duncan test.
2 = ammonium nitrate
Maximum production = 45.2 ton/ha/year (Indonesia at Paquera with 360 kg N/ha/year).
Minimum production = 6.5 ton/ha/year (Criolla at Puriscal with 180 kg N/ha/year).

In Paquera, despite the long dry period and less fertile soils, average yields for all three varieties were almost double those of Coronado, where it rains all year. Edible biomass yield differences were smaller (Table 38). In Paquera lower leaf:stem proportion was observed (Tables 39 and 40) at similar cutting frequencies. This indicates that more frequent cuts can be given in Paquera conditions.

Table 38. Site and nitrogen fertiliser effect on edible dry matter percentage of three Mulberry varieties in Costa Rica

Parameter	Variety			
	Criolla	Tigriada	Indonesia	Average[1]
Site:				
Puriscal	53.8	50.8	38.6	47.7[b]
Coronado	64.7	56.4	48.5	56.5[a]
Paquera	40.0	42.0	32.0	38.0[c]
N, kg/ha/year				
180	52.8	51.5	41.3	48.5[a]
360	52.8	49.1	38.6	46.8[a]
540	52.9	48.7	39.6	47.1[a]
Average[1]	52.8[a]	49.7[b]	39.7[c]	

1 = Values with same letter horizontally do not differ statistically, p<0,05

Table 39. Site and nitrogen fertiliser effect on leaf:stem relation in three Mulberry varieties in Costa Rica

Parameter	Variety			
	Criolla	Tigriada	Indonesia	Average[1]
Site:				
Puriscal	1.00	0.56	0.89	0.82[b]
Coronado	1.40	0.71	0.87	0.99[a]
Paquera	0.61	0.44	0.67	0.57[c]
N, kg/ha/year				
180	1.03	0.60	0.87	0.83[a]
360	0.99	0.54	0.81	0.78[b]
540	1.00	0.57	0.75	0.77[b]
Average[1]	1.00[a]	0.57[c]	0.81[b]	

1 = Values with same letter horizontally do not differ statistically, p<0,05.

Table 40. Site and nitrogen fertiliser effect on growth rates (ton DM/ha/year) of three Mulberry varieties in Costa Rica

Parameter	Variety			
	Criolla	Tigriada	Indonesia	Average
Site:				
Puriscal[1]	92.5	130.0	158.3	126.9[b]
Coronado	74.2	162.5	150.0	128.9[b]
Paquera	186.7	265.8	326.7	259.7[a]
Fertilization[2]				
180 kg N/ha	93.3	150.0	160.0	134.4[b]
360 "	114.2	190.0	235.8	180.0[a]
540 "	145.0	219.2	239.2	201.3[a]
Average	117.8[c]	186.1[b]	211.7[a]	

1 = Values with same letter horizontally do not differ statistically, p<0,05.

12

Mulberry Cultivation and Silkworm Rearing

India has the unique distinction of producing all the four types of silk vis., mulberry, muga, tasar and eri. About 14% of the global raw silk is produced in India, of which more than 65% is mulberry silk, mostly produced in Karnataka. Besides meeting the ever growing domestic demand, silk in its various forms is also exported and this sector is one of the important foreign exchange earner for the country. Of the factors responsible for increasing the production of silk the quality of mulberry plays the most important role.

Soil and Climate

Mulberry is a perennial plant and once established in the field, continues to produce in full form for at least 15 years. Though arboreal in nature, it is trained as low bush for commercial exploitation. Mulberry is a hardy plant and can tolerate varied range of agro-climatic conditions. However, the most suitable range of conditions include a temperature range of 20-300C with sunshine of 9.00 to 13.00 hours per day.

On an average, mulberry requires 50 to 60 mm of water once in a week. Under such conditions (prevailing in southern tropics of India), 5-6 crops can be harvested in a year, whereas in temperate conditions, 2-3 crops are harvested. As mulberry is a deep-rooted plant, the soil for of mulberry garden should be fertile, deep, friable, sandy loam to clayey loam in texture and porous with good water holding capacity. Soils with a pH range of 6.5 to 7.5 are ideal.

Land

Generally flat lands are suitable for irrigated mulberry cultivation. If the slope is more than 15%, suitable land development measures such as contour bunding, bench terracing, etc. should be adopted. Land should be prepared by deep ploughing upto a depth of 30-45 cm in order to loosen the soil and thereafter with a country plough or tractor to bring the soil to a fine tilth. Weeds, stones should be removed during the preparatory stage. Basal dose of farm yard manure @ 20 tonnes per hectare is recommended, which has to be thoroughly incorporated into the soil.

Farm lands along the highways, in the vicinity of factories, area irrigated by raw sewage and untreated effluents, plots abetting other gardens with intensive pesticide usage, water logged areas and tobacco grown land are unsuitable locations for mulberry cultivation.

Variety Selection

The criteria for selection of variety includes fertility of the land, water availability, region specificity, extent of garden and problematic soils. The yield potential of high yielding varieties can be best realised in high fertility soils and they respond to intensive irrigation. Wherever the sise of operation is bigger, it is advisable to have a separate chawki garden for which a variety like S-36 can be chosen in addition to a high yielding variety for rearing late age worms (M-5, S-36 & V-1). In alkaline soils, the variety AR-12 is suitable.

Materials for Planting

Plantation of mulberry garden can be taken up both by cuttings and saplings. However, saplings are always better than cuttings as a planting material in respect of quick and better establishment. Two cuttings / one sapling are planted at each spot in desired spacing, Cuttings are planted keeping one bud exposed while the saplings are planted in pits opened at the spot.

Cutting Preparation

Cuttings should be prepared from well-matured 6-8 months old shoots of about 1.5 cm diameter. Cuttings of 15-20 cm length with 3-4 healthy buds are selected for plantation.

Sapling Raising

Nursery beds with the dimension of 5 m X 1.5 m are prepared. The land is dug to a depth of 30-40 cm and the soil should be pulverised well. FYM is applied @ 15 kg per bed and mixed well with the soil. Some quantity of sand is also added to heavy (clayey) soils to make the soil loose and friable, while some quantity of tank silt or other well decomposed organic matter has to be applied to sandy soils to increase the water holding capacity. Cuttings should be planted in nursery beds with a spacing of 15 cmX10 cm. The full length of cutting is pushed into soil keeping only one bud exposed above the ground. It should be irrigated twice a week. Fertiliser is applied kilogram per hectare after 5-6 weeks when plants sprout well and the root system is established. Saplings of about 80-90 days old can be used for planting. While uprooting the saplings, maximum care should be taken to avoid damage to the root system.

Spacing

Spacing depends upon the soil conditions, slope, variety and convenience for inter-cultural operations as detailed below:

Spacing (Cm.)	Suitability	No.of plants / ha.
90 x 90	Heavy soils, black soils, heavy rainfall areas	12345
120 x 60	Slopy land (hilly areas)	13888
(90 + 150) x 60	Suitable for mechanised farming	13888

Planting

Pit system is ideal in 90 x 90 cm spacing. Pits of 35cm3 are dug and left for one month, which are later filled with FYM and soil at 1: 2 ratio. Trenches of 35 cm x 35 cm are also cent convenient to take up plantation. It is preferable to start the plantation during the rainy season. In hilly areas, 3 cuttings of 20-22 cm length are planted in each pit at a spacing of 15 cms. Cuttings not sprouting in 4-5 weeks need to be replaced by saplings to ensure the required plant density.

Pruning and Training

In order to get good leaf yield for 15-20 years, it is necessary to give a good and sturdy frame to the plant. After the planting, the saplings need to be cut at 10-15 cms above the ground level within a week of planting. After the growth period of six months, the plants are to be cut at a height of 25cm above the ground level keeping 3-4 strong branches.

In leaf picking method, the first bottom pruning has to be taken up at a height of 20-25 cm from ground level and second bottom pruning is done (after 3 leaf harvests) again at 30 cm height from ground level. Afterwards, in leaf harvest method, the middle pruning at a height of 60-70 cm has to be followed once in a year during Jan-Feb and in shoot harvesting method, each shoot harvest done at 25 cm height which itself acts as pruning.

Manure and Fertilizer

FYM has to be applied at 20 tonnes/ ha./ year in two doses following the first bottom pruning and third pruning. Fertilisers has to be applied as per the recommended schedule and secondary and micro nutrients have to be applied wherever necessary. Foliar sprays such as Boron (1%) Urea (0.5%), Zinc Sulphate (0.1%), etc. will improve the leaf quality. The recommended fertliser dose (Kg/ha) is indicated below:

		N	P	K
Rainfed				
	1st yr	50	25	25
	II yr onwards	100	50	50
Irrigated				
Shoot harvest	I yr	100	50	50
	II yr onwards	300	120	120
Leaf Harvest	I year	100	50	50
	II yr onwards	300	120	120

Irrigation

Among the various inputs, irrigation ranks high in giving quick and good results. Regular irrigation at an interval of 8-10 days is ideal. Normally, 1.5 to 2 acre inch of water per irrigation is sufficient. In case of water scarcity, drip irrigation can also be adopted.

Yield

Generally the expected annual yield is 40-50 tonnes of leaves per acre. In shoot harvest method, harvesting can be done at an interval of 70 days (5 harvests). In case of leaf harvest method, first harvest is taken 70 days after 1st pruning and 2nd and 3rd harvest (coinciding with 2nd bottom pruning) at an interval of 55 days. The fourth harvest is taken 70 days after the 3rd harvest and 5th and 6th at an interval of 55 days.

Rearing

The silkworm (Bombyx mori) is the chief source for production of fabulous mulberry silk in sericulture industry. Throughout its rearing period it has to be looked after carefully and nourished with good quality mulberry leaves. Provision for optimum environmental conditions and protection from pests and diseases are the other essential requirements. Depending on the number of generations per year, the silkworms are classified into univoltine, bivoltine or multivoltine breeds. Univoltine and bivoltines breeds are specific for temperate areas whereas multivoltines are for tropical areas. Silkworm rearing involves both technical and commercial aspects. In tropical areas, where mulberry grows luxuriantly throughout the year, 5-6 rearings are conducted in a year while in temperate/sub-tropical areas it is common to raise one/two crops.

From hatching to its full grown stage, the silk worms pass through five instars and the worms up to stage three are called young age worms or Chawki. As they are susceptible to infections and vulnerable to adverse weather conditions, special care is required for rearing of chawkis. The various stages involved in chawki rearing are indicated below:

— *Obtain quality eggs* : Disease free DFLs to be used.

— *Incubation & Brushing* : Ideal storage conditions of 25 to 26°C. temperature and 85% RH, 16 hours of light and 8 hours of darkness ensures uniform hatching (normally between 9-12 days egg laying). Brush newly hatched larvae on to the rearing beds and sprinkle freshly chopped tender leaves of 0.5 to 1 cm sq.

— *Feed the right quality leaf* : Highly nutritious leaves in adequate quantities at suitable intervals of time to be fed. Top 3-4 full grown leaves immediately below the growing buds are suitable.

— Provide right Environment : 26-28°C. and 80-85% RH

— *Feed Management*: Fresh chopped leaves are spread thinly on the dry beds

— Keeping the Beds Clean

— *Care during Moulting*: Provide with optimum temperature and humidity conditions and also necessary spacing and leaf feed. There has to be low to low or no feeding during moulting and maintain a temperature of 24-25°C and RH of 65-70%. The worms will attain maximum growth in minimum time and will take 3 to 3 ½ in the first instar and 2 ½ to 3 days in the second instar to go into moult. The third instar is crucial as this is the intermediary stage between the chawki and late age worms.

Worms begins from third instar. These worms are voracious feeders. The various stages for rearing are indicated in the next page :

— Ensure disinfection two days before brushing of worms. The rearing room and equipments should be disinfected.

— *Provide sufficient quantity of leaf.* The late age stages consumes 95% of the total feed. Mulberry leaves to be harvested with optimum maturity with relatively low moisture, high proteins and fibres with less carbohydrate content.

— *Ensure good environment*: Maintain 24-25o C temperature and 70-75% humidity. The space requirement for 100 DFLs - for multi-voltine 100-200 sft and 200-400 sft. space for 4th and 5th instar, respectively : For bi-voltine 115 to 225 sft and 225 to 550 sft. space for 4th and 5th instar, respectively.

— Ensure hygenic conditions

— Protect from natural enemies like Uzi fly

— *Mounting of ripe worms*: Ripe worms to be picked on completion of 6-7 days after passing 4th moult and provide 26-27°C and 55-60% RH to ripe worms.

— *Harvesting of Cocoons*: Suitable mountages are used for mounting optimum density of worms (40-45 worms/ sft.) are maintained. Mountages with worms not to be exposed to direct sun light and humid conditions to be avoided. Cocoons harvested on 5th to 6th day after mounting which ensures complete development of pupae. The qualities of good cocoons include uniform size and shape, rich silk content, less floss, more shell weight, uniform shape, uniform build of cocoons.

— *Cocoon Yield*: The average yield is 40-45 kg. (CB) and 45-50 kg (BV) from 100 DFLs.

The entire process of silk worm rearing from egg hatching to cocoon takes about 26-27 days.

Rearing House

Mulberry silkworm rearing, being completely domesticated, demands specified environmental conditions like temperature (24-28°C) and relative humidity (70-85%). It is therefore necessary to evolve measures for economic cooling through selection of proper material for wall and roof fabrication, orientation of building, construction method, design, etc. Further, enough space must be available to carry out leaf preservation, chawki rearing, late age rearing and moulting. It should also be convenient enough to conduct effective cleaning and disinfection.

The size of the rearing house depends upon the quantum and type of rearing. The different types of rearing are, shelf or stand rearing, platform rearing and floor rearing. Shelf rearing requires minimum space and is most common in India. A floor area of 150 sft. can provide rearing space for 100 CB DFLs or 75 BV DFLs.

In platform rearing, the larvae are shifted to shoot rearing platforms after 3rd moult. Ideal size of each platform will be 5 x 25 ft. which can accommodate 50 DFLs upto spinning. Platforms can be arranged in two/three tiers with a gap of 2-3 ft. A building floor area of 250-300 sft. is required to rear 100 DFLs of CB layings or 75 DFLs of BV layings.

Equipments

Various equipments required for rearing 100 DFLs are indicated below :

Sl.No.	Particulars	Cross breed (Multi x Bi)	Bivoltine
1	Rearing trays 3 ½ ft. x 4 ft. dia	50 no.	60 no.
2	Rearing shelves (10 tiers)	no. 4	6 no.

Contd....

Contd....

3	Feeding stand	one	one
4	Branch/ leaf cutting knife	one	one
5	Bed cleaning nets	50 no.	60 no.
6	Ant wells	20 no.	24 no.
7	Parafin paper/ Blue polythene sheet	50 m	60 m
8	News paper	8 kg	10 kg
9	Wash basin	one	one
10	Mats (6' x 4' size)	2 no.	2 no.
11	Gunny cloth	20 m	20 m
12	Foot mat	one	one
13	Bamboo baskets	Four	Four
14	Plastic basins	Four	Four
15	v Chandrike (6' x 4')	40 no.	50 no.
	v Collapsible mountage	100 no.	120 no.
	v Rotary mountage		35-40 no.

Investments

The cost worked out for various investments are indicated below:

Sl.No.	Investments	Cost (Rs.)
1	Mulberry development	24000/ha.
2	Rearing equipments	
	Shoot rearing	24,000 - 2 ac. model 44,000 - 4 ac. model
	Leaf feeding	16,000 - 1 ac. model
3	Isolation chambers (10 Nos.)	8,750/ha.
4	Rearing house (leaf feeding method)	
	i) RCC/ Stone slab roof	
	200 sq.ft.	30,000
	375 sq.ft.	52,000
	ii) Mangalore tile/ ACC sheet roof	
	200 sq.ft.	24,000
	375 sq.ft.	47,000

Contd....

Contd....

5	Shoot Rearing hall/house (1000 sq.ft.)	
	i) RCC/ Stone slab 1000 sq.ft.	246,000
	ii) Mangalore tile/ ACC sheet roof	180,000
6	Reeling units	
	i) Cottage basin	130,000
	ii) Improved Charkas	54,500

Loan repayment

The term loan repayment by the ultimate borrower will start based on the annual net surplus available from the investment made. The general repayment program suggested for various activities in Karnataka are indicated below:

Sl.No.	Activity	Maximum repayment	General grace
1	Mulberry	4	11
2	Rearing equipment's	3	11
3	Isolation chambers	3	11
4	Rearing house		
	i) Conventional rearing	5	11
	ii) Shoot rearing	7	
5	Reeling unit		
	i) Cottage basin	7	11
	ii) Improved charkas	5	11
6	Irrigation investments (Ground water irrigation structures, SIP sets, drip etc.)	9	11

13

Regional Sericulture

Sericulture in Orissa is an agro based cottage industry. It is popular among rural households. SC/ST and rural households like commercially weakest section of the society practicing this culture to earn their livelihood.

Sericulture in Orissa includes both Mulberry and Vanyasilk. Tassar and Eri silk production comes under Vanya Silk. Tassar & Eri culture is traditional to the state, whereas Mulberry is nontraditional.

More than 40,000 Tassar rearers have organised 62 primary Tassar Rearers cooperative society. 15989 hectares of natural Assan & Arjun plantation and 6830 hectares of systematic Assan & Arjun block plantation were the natural resources to sustain this traditional culture in the state. Due to massive deforestation our natural wealth declined to 7620 hectares for Tassar silkworm rearing.

15 numbers of Pilot Project Centers have been setup in Tassar growing areas to cater the Basic seed demand of the seed rearers. To popularise the industry the CSB under Government of India has setup 4 Basic seed Multiplication & Training Centers (BSMTC) for quality seed supply and one REC to transfer technologies from lab to land.

Eri culture is practiced as a subsidiary occupation in limited pockets of the state, where castor is grown as oil seed crop. About 500 numbers of Eri farmers are actively involved in Eri culture, getting support through 12 numbers of Eri Centers and one Eri Rearer Co-operative Society. Three numbers of Government Eri Seed farmers have been setup to meet the requirement of commercial Eri silkworm rearers.

Moriculture and Mulberry silkworm rearing is nontraditional to the state. seed farms and grainages have been established to provide backward linkage to the farmers. Mulberry Extension Service Centers (MESC) at Gram Panchayat level have been organised to provide inputs support and preparation of chawkiworms for subsequent supply to farmers of commercial crop.

Production statistics 2004-05

Sl. No.	Sector	Dfls Production /Utilisation	Cocoon Production		Yarn production	
			Qty.	Value (Rs. in Lakh)	Qty.	Value (Rs. in Lakh)
1	2	3	4	5	6	7
1	Tassar	957038	K26229	208	32.5	260
2	Eri	18210	Kg.1177	1.6	0.58	18.45
3	Mulberry	74706	Kg.18457	12.9	1.84	4.36

The Mulberry Rearers Co-operative Societies (MRCS) have been organised by the farmers at Block level, which is giving forward linkage to the farmers. For value addition 4 numbers of reeling infrastructures have been set up for utilisation of farmers produce (Cocoon) and production of Raw Silk. This Raw Silk serves as the raw material in the handloom sector in the state.

Government of Orissa is trying to stabilise the sector by introducing Mulberry varieties like, V1, S1 & C1730 under irrigated condition and high yielding variety of silkworm breeds like CSR hybrids. The Mulberry rearers of Orissa have started taking up CSR hybrids in two favorable season and three crops with cross breed (Multivoltine X Bivoltine) in unfavorable seasons. About 1500 farmers of the state are closely associated in Mulberry silkworm Rearing.

Vision

Sericulture

— Planned to produce 82,000 basic DFLs for self-dependency.

— Mobilisation of working capital from CSB (Rs. 60.00 Lacs), welfare department. (Rs.20.00 Lacs), KBK programme (10.28 lacs) a.

— Tasar rearing activities are accepted as forestry activity.

— Government policy for tasar plantation in clusters under Joint forest management programme.

— Special Project proposed to tap the utilisation of forest and human resources.

Strategy

— Self-employment opportunity to the tribal/BPL families.

— Plantation of trees like Assan & Arjun for utilisation in silkworm rearing and production of Tassar Cocoons.

— Up-gradation of degraded forest land for Tassar culture

— Mulberry silkworm activity in in-door getting leaf from mulberry garden providing periodical income to the farmers alround the year.

— Growing castor plants in wild or cultivable farms for silkworm rearing to provide additional income to the castor growers through Eri culture.

Achievement

The object of the scheme is to provide assistance to the Sericulturist by supply of planting materials, disinfecting material, preparation of Chowki worms, incentive bonus to bi-voltine rearers, study tour & training to rearers, organisation of seminars, farmers-day, exhibition, Chowki rearing training, printing of publicity material etc. The schemes are being implemented by assistance from the state Govt. and Govt. of India through Central Silk Board. During the year 2004-05, a sum of Rs.30.00lakhs has been spent towards state share of assistance.

During the 10th five-year plan 132.28 metric tone of yarn was produced providing employment opportunity to 17340 farmers by end of 2004-05.

During the year 2004-05, 32.5 M.T. of tasar silk, 1.85 M.T. of mulberry silk and 0.55 M.T. of eri silk was produced and in turn 13508, 2935 and 896 persons were employed in three sectors respectively.

Promotion of Sericulture Industries continued to operate in the state along with KBK districts under RLTAP. During 2004-05 a sum of Rs.50.00 lakh were sanctioned and 295 families were benefited in mulberry plantation and silkworm rearing. Under the infrastructure development programme, one MESC, one MRCS, one Eri farm-cum-grainage house and renovation of two Tassar seed infrastructures were taken up during the year to support the beneficiaries.

Action Plan

The assistance for the following components have been proposed under RLTAP for KBK district during 2005-06.

Tassar programme

— Assistance to tassar seed rearers for maintaining host plants.

— Assistance to commercial rearers for maintaining chowki garden.

— Rearing equipment support to seed rearers.

— Support to commercial rearers for rearing e.quipment

— Support to private tassar graineurs for acquiring Grainage equipments and construction of a grainure.

Mulberry programme

— Support for raising nursery with high yielding mulberry varieties.

— Assistance to farmers for construction of rearing houses.

— Support for preparation of Vermi composts.

— Institutional supply of quality disinfectants.

— Assistance to the beneficiaries for mulberry plantation, rearing house, rearing equipment and fencing.

— Establishment of mulberry extension service center.

— Maintenance of infrastructure

— Support for providing Irrigation to mulberry plantation.

Eri Programme

- — Augmentation of Eri food plants with training and start-up tools.
- — Assistance for construction of Eri rearing house.
- — Support for construction of Eri farm-cum-grainage house and acquiring required equipments.

During the year 2005-06 a sum of Rs.50.00 lakh have been provided with target of benefiting 47 beneficiaries in Tassar sector and construction of 3 private tassar graineurs in KBK district under RLTAP including construction 208 rearing houses for 208 beneficiaries in mulberry sector, preparation of Vermi composts for 26 units and supply of quality disinfectants to 113 farmers.

It is proposed to provide assistance for Construction of 208 nos. of Eri rearing houses to Eri farmers and under infrastructure development programme one Eri farm-cum-grainage house is proposed to be renovated in KBK districts.

The annual plan for the State during 2005-06 in all the three sectors inclusive of KBK is planed as per the following details.

Sl. No	Particulars	Mulberry	Tasar	Eri
1	Food Plantation	240Acre. (Mulberry)	1740 Plant Hectare (Assan & Arjun)	545 Acre (Castor)
2	DFLs to be brushed	1.75 Lakh nos.	16.5 lakh nos.	1.09 lakh nos.
3	Cocoon to be produced	66 M.T.	51600 Kahans	7.6 M.T.
4	Nos. of Farmers to be benefited	1450	7348	1090

Schemes

Plan Schemes (State Plan)

A number of state plan schemes are in operation to provide necessary input assistance to Government farms /units and beneficiaries for development of Sericulture in the state. They are as follows.

a. Promotion of sericulture industry:

The scheme provides assistance to units and beneficiaries for development of the sericulture in the state through the following sub schemes.

i) Seed multiplication center:

The objective of the scheme is to provide operational assistance to different Tassar seed multiplication centers established in the state for production of commercial DFLs.

ii) Mulberry cuttings:

Under the scheme high yielding variety mulberry cuttings are supplied to the farmers at free of cost for raising new plantation or substitution of old varieties in the existing mulberry acreage.

iii) Progressive farmers:

The objective is to encourage progressive farmers for taking up mulberry sericulture. Assistance for Rs.5000 against the provision of Rs.20, 000/- is given to progressive farmers for construction of a house suitable for Mulberry silkworm rearing.

iv) Mulberry extension service centers:

MESCs are established at Gram Panchayat level of the

state to produce quality chawki worms. These centers provide technical support like technical know-how to the farmers at doorstep. The existing chowki gardens and chowki rearings are maintained with the assistance under the scheme.

v) Preparation and distribution of Chawki worms:

The required chowki worms are reared and maintained through these units under healthy condition and supplied to the farmers after 3rd moult for raising healthy cocoon crop. The expenditure under this programme is met out of this scheme.'

vi) Study tour and training:

The objective of the scheme is to provide training to the farmers, Government officials, bankers etc., under sericulture technology.

vii) Publicity and Propaganda:

Camps, meetings, exhibitions, workshops are organised to popularise the new technologies among the sericulturists.

viii) Incentive Bonus:

Incentive is released to the mulberry Bi-voltine silk reelers for production of quality BV silk.

ix) Mass Disinfections:

Disinfecting materials are purchased and supplied to the farmers for disinfections of rearing house and equipments for prevention of outbreak of silkworm diseases.

b. Support to SERIFED for development of Sericulture:-

There is fluctuation in raw-silk rate in the National and

Inter-national market. The mulberry cocoon price is also going through ups and downs. The tribal rearers of the state are unable to absorb such shock. The objective of the scheme is to create price fluctuation /stabilisation fund at the SERIFED level to overcome such hardship.

c. Share Capital Investment:-

The objective of the scheme is to provide Government share capital to the Apex co-operative for smooth marketing of the produce of Tassar, Mulberry & Eri growers and in ensuring timely payment of cocoon value.

d. Intensive development of Sericulture in KBK Districts:-

The sericulture programme has been taken up widely in the rural areas of KBK districts as it provides ample scope for better earning of the rural people. It is proposed to take up mulberry sericulture in Koraput, Rayagada & Kalahandi districts for the development of the rural people. The Tasar programme has also been taken up in Nuapada district to provide employment to the tribal people by way of Tasar silk-worm rearing and production of Tasar cocoons. The eri plantation and eri-silk worm rearing has also been taken up in Rayagada and Kalahandi districts. The objective is to provide self-employment to the interested farmers of these areas under the ungoing schemes of Tassar, Mulberry & Eri.

Catalytic Development Schemes

Central silk Board has been implementing central sector and centrally sponsored schemes for the development of sericulture with the involvement of the State Government. These schemes have been collectively prescribed as Catalytic Development Programme (CDP). Some of the problems faced in implementation of the programme during IX plan have been addressed before finalising the programme for the X plan period, while the overall objective of the programme is to ensure coordinated effort to support sericulture at all stages of value addition.

Mulberry Sector

1. Support for Raising Nurseries of High Yielding Mulberry Varities

The objective of the scheme is to help State Government to produce saplings of high yielding mulberry verities from Kissan nurseries raised by farmers for supply to new farmers and sericulturists who wish to re-plant their farm with the improved varieties.

The unit cost of Rs. 0.50 per sapling is paid to farmers from whom the saplings are procured.

Cost Sharing

CSB	State	Beneficiary
50%	50%	—

2. On-farm Training and Start-up Tools to New Mulberry Sericulturists

— The objective of the scheme is to help farmers to take up sericulture on scientific lines and equip them with essential tools /equipments to embark on mulberry farming and rearing of silkworm & to improve quality and productivity of cocoons through adoption of scientific packages of practices in mulberry cultivation and silk worm rearing.

— The unit cost is Rs. 2000 per acre – Rs. 1750/- for supply of tool kits and Rs. 250/- for providing stipend to farmers for one month training (if area of mulberry is less than one acre, cost of tool kit would be Rs. 900/- and Rs. 250/- stipend for one month training).

— The entire cost will be born by the Central Silk Board.

— Farmers identified by the State Department of Sericulture are eligible for this assistance.

3. Rearing Appliances and Farm Equipment to Farmers (For bivoltine silkwom rearing)

— The objective of the scheme is to assist the farmers in purchasing modern tools and equipment which make mulberry cultivation and rearing more efficient and convenient & to encourage more farmers to take up Bivoltine sericulture successfully with the help of modern tools, machinery and equipment.

— The unit cost is Rs. 30,000/- per farmer (for a minimum of 1 acre plantation)

— Farmer who has sufficient experience in silkworm rearing and are willing to take up bivoltine silkworm rearing are eligible for this assistance.

Cost Sharing

CSB	State	Beneficiary
25%	25%	50%

List of Rearing Appliances /Farm Equipments

A) General Farmers

Sl. No.	Item
1	Rearing Trays (Plastic)
2	Incubation Frame
3	Nylon Net
4	Bed Disinfectant
5	Paraffin Paper
6	Hygrometer
7	Egg carrying bag / box
8	Duster

A) Bivoltine Farmers

Sl. No.	Item
1	Rotary mountages
2	Shoot rearing rack
3	Nylon nets
4	Pruning saw
5	Secateur

4. Assistance for Intallation of Drip Irrigation System

— The objective of the scheme is to assist mulberry farmers in conserving water resources which are scarce in most parts of the country so that farmers would be able to take up farming more efficiently and get increased output of quality mulberry leaf.

— The unit cost is up to maximum of Rs. 48,000/- per hectare.

— Farmers, who won land, have an assured source of water and power connection is eligible for the assistance.

Cost Sharing

CSB	State	Beneficiary
45%	5%	50%

5. Support for Use of Quality Disinfectants

— The objective of the scheme is to encourage the farmers to use quality disinfectants and prevent crop losses due to diseases.

— The unit cost is Rs. 2870/- per farmer per year. If number of silkworm crops are only three or less in a year, Rs. 500/- per crop.

— Both bivoltine silkworm rearers and rearers of other silkworm races / hybrid are eligible. Seed rearers adopted by Central Silk Board units are also eligible for the subsidy.

Cost Sharing

Category	CSB	State	Benefeciary
F1 Farmers	25%	25%	50%
P1 Farmers	50%	50%	——

Disinfection modules

Sl. No.	Item	Quantity
1	Bleaching Powder	8.790 Kg.
2	Slaked Lime	2.171 Kg.
3	Uzi trap	10 nos.
4	Sanitech (Chlorine Dioxide)	4 lts.
5	Bed disinfectant	5.000 Kg.

6. Assistance to farmers for construction of rearing house

— The objective of the scheme is to assist farmers to acquire separate rearing houses, which contributes to improvement in quality and productivity of cocoons. Production of cocoons would take place in hygienic conditions and expenditure on disinfectants and losses on account of diseases and pests would be minimised.

— The unit cost depends on the size of the rearing house, up to a maximum unit coast of Rs. 1,00,000/- per rearing house.

— Farmers who won mulberry gardens, site / land for construction of rearing houses. Farmers should also have an experience of at least 2 years in sericulture practices.

Cost Sharing

CSB	State	Beneficiary
25%	25%	50%

7. Chowki Rearing Centres—Assistance to Quality Service Clubs/Non-government Organisations/Self-help Groups for Construction of Chawki Rearing Building and Procurement of Chawki Rearing Equipment

— The objective of the scheme is to ensure the success of cocoon crops through professional and scientific management of young age silkworm rearing (Chawkies).

— The unit cost is Rs.1, 00,000/- for construction of a chawki rearing building and Rs. 2,00,000/- for purchase of chawki rearing equipment.

— Quality Service Clubs / NGOs / Self – Help Groups identified and recommended by the State are eligible. Quality Service Clubs should have at least 20 members.

Cost Sharing

CSB	State	Beneficiary
25%	25%	50%

List of Equipments for Chawki Rearing.Center

Sl. No.	Item
1	Microscope
2	Plastic Rearing Tray (Size 3'x4')
3	Temperature & Humidity Recorder
4	Heater with thermostat
5	Humidifier
6	Power Sprayer
7	Mask
8	Incubation frames (Plastic)
9	Leaf collecting baskets
10	Chopping board + knives
11	Cleaning nets for chawki
12	Bottom stands
13	Feeding stands
14	Leaf chopping machine
15	Flame gun + cylinder
16	Hand duster
17	Litter basket

18	Plastic basins
19	Wash basins
20	Secateurs
21	Egg carrying box / bag
22	Ant wells
23	Foot mats

8. Crop Insurance Support—Mulberry

— The objective of the scheme is to cover the risk of crop losses faced by farmers who have not had much experience in rearing Bivoltine silkworms.

— A premium of Rs. 144/- for 100 bivoltine Dfls (Rs. 180/- per 100 Dfls in respect of Andhra Pradesh) and Rs. 31/- per beneficiary towards personal accident to the rearer.

— All bivoltine silkworm rearers are eligible. While farmers rearing silkworms other than bivoltine may also go for crop insurance, subsidy is presently available only to bivoltine silkworm rearers.

Cost Sharing

CSB	State	Beneficiary
40%	40%	20%

9. Support for Uprading Equipment in Existing State/Parastatal/Co-operative owned seed Production and Reeling Units:

— The objective of the scheme is to improve the infrastructure available with the State / Co-operatives so as to achieve improvement in quality and productivity in seed production and reeling.

— The unit cost depends on the nature of up gradation required.

— States / Parastatal / Co-operatives are eligible for this assistance.

Cost Sharing

CSB	State	Beneficiary
50%	50%	—

List of Equipments for up-gradation of infrastructure in seed production and reeling units

Sl. No.	Name of the Equipment
	Silkworm Seed Production Centre
	A. General Equipments
1	Cocoon /Pupae preservation plastic trays
2	Plastic oviposition trays
3	Plastic seed cocoon transportation crates
4	Plastic male moth preservation trays
5	Room heaters
6	Humidifier /Air coolers
7	Egg preservation cabinet
	B. Machineries
1	Electric balance
2	Power sprayer
3	Generator (Mini)
4	Incubators
5	Platform weighing balance
	C. Testing & Treatment Equipment
1	Moth examination table
2	Moth crushing Mixi
3	Centrifuge
4	Cyclomixer
5	Microscopes
6	Deflossing Machine
7	Loose egg preparation table
8	Acid treatment bath
	Reeling Units
1	Cocoon driers (Hot air drier or ushnakot)
2	Cocoon sorting table
3	Cocoon storage racks

4	Circular pressurised cooking machine
5	Two pan cooking unit
6	Cocoon cooking cages and pressing disk
7	Multiend Reeling Machine
8	Re-reeling machine (closed type)
9	Silk examination stand
10	Small skeining machine
11	Long Skeining (book) machine
12	Long skein stand
13	Small skein book press
14	Small reel permeation chamber
15	Denier scale
16	Single cocoon reeling epprouvette.
17	Balances – 2 (one 0.001 grams to 100 grams and one 5 kg. cap
18	Boiler
19	Water softening soil plant

Tassar Sector

1. Assistance for Raising and Maintaining Systematic Plantations of Tassar Food Food Plants (Assan and Arjuna)

— The objective of the scheme is to improve the quality and productivity of Tassar and oak Tassar food plants by adoption of technologies developed by Research Institutes of Central Silk Board.

— The unit cost is Rs. 3900/- per hectare for maintenance of systematic plantation of Tassar host plants, Rs. 800/- per farmer for development of chawki garden and Rs. 10,000/- per hectare for raising plantation of Oak.

— Members of Vana Samrakshana Samithis, Co-operatives, Self-Help Groups, Rearers' Groups sponsored by NGOs / Individual Rearers etc., identified by the States.

Post Cocoon Sector

1. Support for setting up Multiend Silk Reeling Units

— The objective of the scheme is to improve the quality of yarn and also modernise the reeling sector. Since Multiend reeling machines are capable of producing high quality silk, production of quality cocoons also will get a boost. Weaver would be able to source warp quality yarn for power looms or handlooms.

— The unit cost of the scheme is Rs. 5.45 lakhs, Rs.7.22 lakhs and Rs. 14.00 lakhs for 6 basin, 10 basin, and 20 basin units, respectively.

i) For 6 basin and 10 basin units - 50% subsidy by CSB & State

ii) For 20 basin unit - 35% subsidy by CSB. State govts. have the option to the subsidy to the extent they can

Cost Sharing

Category	CSB	State	Benefeciary
6 Basin or 10 Basin	35%	15%	50%
20 Basin	35%	Optional	65%

Support for Setting up Multiend Silk Reeling Units

Multiend Reeling Package with 6 Reeling basins	
1. Hot Air drier 50Kg. Capacity or	
Ushnakothi – 50 Kg. Capacity	1 No.
	1 No.
2. Cocoon sorting table	1 No.
3. Circular pressurised cooking machine	1 No.
4. Two pan Table for brushing	2 Sets
5. 6 Basin capacity Multiend Reeling Machine (10 end per basin)	1 No.
6. 6 window capacity Re-reeling machine	1 No.

Contd.....

Contd.....

7. Silk Reel permeation chamber	1 No.
8. Long Skein Silk Book making machine	1 No.
9. Silk testing appliances – Denier scale and epprouvettes	1 Set
B. Multiend Reeling Package with 10 Reeling Basins	
1. Hot Air drier 50Kg. Capacity or Uhsnakothi – 50 Kg. Capacity	1 No. 1 No.
2. Cocoon sorting table	1 No.
3. Circular pressurised cooking machine	1 No.
4. Two pan Table for brushing	4 Sets
5. 10 Basin capacity Multiend Reeling Machine (10 end per basin)	1 No.
6. 10 window capacity Re-reeling machine	1 No.
7. Silk Reel permeation chamber	1 No.
8. Long skein Silk Book making machine	1 No.
9. Silk testing appliances – Denier scale and epprouvettes	1 Set
C. Multiend Reeling Package with 20 Reeling Basin	
1. Hot Air drier 50Kg. Capacity or Ushnakothi – 50 Kg. Capacity	2 Nos. 2 Nos.
2. Cocoon sorting table	1 No.
3. Circular pressurized cooking machine	2 Nos.
4. Two pan Table for brushing	8 Sets
5. 6 Basin capacity Multiend Reeling Machine (10 end per basin)	2 Nos.
6. 6 window capacity Re-reeling machine	2 Nos.
7. Silk Reel permeation chamber	1 No.
8. Long skein Silk Book making machine	1 No.
9. Silk testing appliances – Denier scale and epprouvettes	1 Set
D. Steam Generating system (Boiler) of 100 kg Steam output / hour	
E. Steam Generating system (Boiler) of 200 kg Steam output / hour	

— All reelers who have not availed the subsidy during IX Plan under the Multiend reeling scheme and are interested to set up new units are eligible for this assistance.

2. Support to Silk Reeling Units

A. Interest Subsidy on working capital loan sanctioned by Banks to Reeling Units :

— The objective of the scheme is to encourage flow of working capital to new Multi-end reeling units. The scheme will also reduce the interest burden on working capital loan raised from Banks and encourage entrepreneurs to invest their funds in modern Multiend reeling units.

— The extent of assistance provided is as follows:

Mulberry Reeling Units -	Rs.3.00 Lakhs **
Tassar Reeling Units -	Rs. 1.50 Lakhs
Tassar / Eri Spinning Unit -	Rs. 0.50 Lakh

** Applicable for 10 basins and above. For 6 basins, R. 2.00 lakhs will be provided.

— The amount indicated above are the amounts that CSB would keep as deposits in the Bank / State Financial Corporation which extend working capital to multi-end reeling units.

— All multi-end reeling units in Mulberry sector and reelers / spinners in the Non-mulberry sector are eligible for this assistance.

B. Incentive for production of Bivoltine Silk on Multiend Reeling Machines.

— The objective of the Scheme is to encourage to produce more bivoltine silk and help them to compete with cheap imports.

— The unit cost is Rs. 100/- per Kg. of bivoltine silk.

— All reelers of Bivoltine silk are eligible to get the assistance.

Cost Sharing

CSB	STATE
50%	50%

Cost Sharing

CSB	State	Beneficiary
25%	25%	50%

Lit of Equipments for Cocoon Quality Assessment

Sl. No.	Equipment
1	2
1.	Electronic Balance (0.001 gms sensitivity, 200 gm capacity)
2	Physical Balance (Two Pan) or Electronic Balance (0.01 gm sensitivity, 5 kg capacity)
3	Cocoon sorting table
4	Cocoon preservation Rack
5	Lab model batch type hot air drier with thermostat arrangement
6	Lab model single pan cooking vessel with electric heating arrangement with witch board
7	Cocoon crates of 2 Kg capacity, 250 ml plastic containers, cotton /nylon net, scientific calculator, weight box for physical balance, stainless steel bowl, digital thermometer, kilcha stand
8	Motorised pirouette
9	Testing table /stools /benches

3. Assistance to Quality Service Clubs for procurement of Cocoon Quality Grading and Raw Silk Testing Equipments.

— The objective of the Scheme is to decentralise and popularise grading of cocoon and yarn among farmer and reeler which would benefit them from quality based pricing.

— The unit cost is Rs. 1.00 lakh for procurement of cocoon quality grading equipments and R. 2.00 lakhs for procurement of raw silk testing equipments.

— The Quality Service Clubs which have atleast 20 members recommended by the State Department would be eligible.

14

Mulberry Dykefish Pond Model

The mulberry dyke fish pond complex is a system developed by farmers in the China's Pearl River Delta region to make full use of available land and water resources. It is an interrelated ecosystem that brings into full play the productive potential of humans and their environment and promotes the development of different branches of agriculture.

The co-production of mulberry growing and sericulture with fish farming, where the banks of the fish ponds were turned into mulberry dykes, was first introduced at around late Ming dynasty (16th Century) in the northern part of the Pearl River Delta, in such places as Jiujiang, Xiqiao, and Datong in Nanhai County, with Jiujiang as the centre. From Nanhai County, the interrelated production of silk and fish began to spread gradually to the middle and southern parts of the delta, to Shunde, Panyu, Zhongshan and Xinhui Counties.

Shunde County in the Pearl River Delta region used to contain the greatest number of fish ponds and practised fully the mulberry dyke fish pond model because of its central location of the delta, its extensive low-lying land and its excellent water transportation. In

1905-1906, for instance, the export of raw silk from Guangdong Province amounted to about 40 million silver dollars, with Shunde alone accounting for three-fourths of this volume. In the 1920s, China's export of raw silk reached its historic peak and the dyked pond agriculture recorded its climax of growth. After 1926, however, the silk trade suffered a severe decline on the international market as Japanese raw silk became a major competitor. The land area planted with mulberry trees in the Pearl River Delta has since decreased rapidly, dropping from about 1.2 million mu (or 200,000 acre; one mu is one-sixth of an acre) at around 1926 to 350,000 mu (about 58,300 acre) in 1978. In Shunde, for example, the area devoted to mulberry dyke fish ponds had dropped from 660,000 mu in 1926 to 190,528 mu in 1978. This practice is now vanishing rapidly due to increasing pressure coming from coastal area development projects for industrialization and urbanization purposes. At present, some of the mulberry dyke fish ponds in the delta have been transformed to cane dyke fish ponds, which are gradually migrating towards the sandy and salty fields further toward the coast. There has been a gradual revival of the mulberry dyke fish ponds or fish ponds of mulberry and cane dykes in alternate years.

The mulberry dyke fish ponds are traditionally managed mulberry growing, sericulture and fish farming. They are build in low-lying land area in the Pearl River Delta and their operation is carefully designed to fully utilize natural resources to maximize production outputs without detriment to plants and animals in the process.

The mulberry dyke fish pond eco-system serves two major functions: (a) achievement of a general ecosystem balance through the harmonization of well-coordinated activities and functions embedded in the ecosystem, and

(b) transformation and regeneration of organic substances based on a multi-layer trophic eco-system structure, which helps contribute successfully to sustainable economic development in the Delta region.

Description of the activity

The mulberry dyke fish pond complex contains two interrelated systems of dyke and pond; the dyke is the land ecosystem for the growth of mulberry trees whereas the pond is the water ecosystem, consisting of fish and aquatic plants. Mulberry leaves are fed to the silkworms, whose excreta are used as fish food, and the fertile pond mud-consisting of fish excreta, organic matter, and chemical elements-is brought up from the bottom and used as manure for the mulberry trees. The mutual benefits of each link in the system have been known to the farmers of the area for many years, as reflected in the folk saying that "the more luxuriant the mulberry trees, the stronger the silkworms and the fatter the fish; the richer the pond, the more fertile the dyke and the more numerous the cocoons." In this system, the mulberry tree represents the first trophic level. Photosynthesis takes places in its leaves, which are fed to the silkworms (the primary consumers) whose excreta and chrysalises are in turn fed to fish (the secondary consumers). The aquatic organisms in the pond are the reduction agents that decompose fish excreta and algae, break down the organic matter in the pond, and produce nitrogen, phosphorus, and potassium. These are then returned to the mulberry dyke when pond mud is used as fertilizer.

The success of the system relies on consolidating the interrelationship between the dyke and the pond. The pond supplies the dyke with the fertile mud that the peasants heap onto the dyke two or three times a year;

with each layer measuring at 5 to 6 cm thick. Pond mud decomposes on the dyke and then becomes the primary source of fertilizer for the mulberry trees. The peasants of Shunde County estimate that 100,000 kg of pond mud are the equivalent of 50 kg of chemical fertilizers, and that each heaping adds the equivalent of 15 kg of urea to the soil. The mud is better than chemical fertilizer because it is composed of about 5 percent organic matter, which the plants need. In addition, the mud acts as a weed killer, retards water evaporation, and maintains soil fertility for a longer period of time.

The effects of the mud on the mulberry trees are particularly noticeable during the dry but still warm months of September and October. During heavy rains some of the mud is washed back into the pond, where it is mixed with the remains of plankton and reacquires its fertilizing qualities. In this way the dykes are constantly supplied with a renewable source of fertilizer. Through photosynthesis the algae and aquatic plants in the pond give off oxygen and produce glucose, added nutrients that benefit both fish and aquatic plants.

Capacity building

The transformation of some of the mulberry dyke fish ponds to cane dyke fish ponds in the delta region has been assisted by agricultural extension services run by local-level governments as well as local research institutions. Both parties have played an important role in helping to sustain a noteworthy degree of awareness among local peasants on the economic benefits and ecological significance of maintaining the interrelated twin eco-systems of dyke and pond in the delta region. The latter has been particularly critical in building and supporting local capacity among the local population to help them adapt their production practices to suit changing environmental and ecological conditions through fieldbased experiments and pilot projects.

Technologies

Innovative agro-eco-techniques have been developed by local agricultural research institutions to help peasants to adapt continuously their production practices to changing ecological and environmental factors, which in turn help maximize the return from their investments in the dyke-pond eco-systems. For instance, the development and introduction of new production techniques in raising pollutionresistant silkworms has led to a gradual revival of the mulberry dyke fish ponds in recent years after a sustained period of decline in the 1990s decade. Moreover, the introduction of another innovative production method-raising fish ponds of mulberry and cane dykes in alternative years-has helped restored the natural productive capacity of the delicate land-water eco-systems in the delta region.

Design, planning and management

Given that the mulberry dyke fish ponds are built in low-lying, ecologically sensitive land areas and that the maximization of their production outputs requires the full utilization of natural resources within the twin land-water eco-system complexes, the discrete and yet inter-linked, multi-layer activities of the operation of these ecosystems have to be carefully planned and managed to achieve a general balance and harmony in the entire ecosystem. The successful operation of these interrelated systems thus demands careful and well-coordinated design, planning and management on the part of the relevant key stakeholders; namely the peasants, local government officials, and scientists based in local agricultural research institutions.

From a broader planning and managerial perspective, while the mulberry dyke fish ponds have been revived and maintained in selected sites in the Delta region, they are no longer a major source of local livelihood per se for most of the peasants. Nevertheless, theses land-water ecosystems and their variants-such as fish ponds cane dykes-are still offering the peasants multiple and supplementary sources of income through the sale of harvested agricultural products as well as the promotion of such sites as popular tourist destinations.

The mulberry dyke fish pond complex is a system in which plants, silkworms and fish live in a mutually dependent and beneficial ecological whole. The system maximizes the use of local resources, generates a variety of products, provides multiple sources of income for the peasants, and creates different kinds of employment opportunities in an area under great population pressure. Because of the warm climate of the delta, mulberry leaves can be harvested in the same year in which the trees are planted. They are picked every 20 days or so, and eight or nine silkworm hatchings are possible each year.

The peasant's income is therefore increased. Multiple fish harvests are also normal. The fry are put into the pond at the beginning of the year, and the fish are netted every 3 or 4 months. As a result of multiple sources of income and the high prices the peasants receive for silk and fish, their income level is much higher than that of those peasants for whom rice or sugar cane is the chief source of income.

Impacts

Traditional mulberry dyke fish pond practice has been

proven to be an environmentally friendly and sustainable sericulture method because it maximizes production on a continual basis, improves water quality and reduces serious flooding in the low-lying land, avoids over-feeding and the misuse of chemical fertilizer, and provides natural food for the silkworms and fish within the interrelated ecosystem.

The peasants of the Pearl River Delta have learned from experience that the amount of silkworm excreta and chrysalises produced by a mu of mulberry trees can meet the food requirements of fish raised in a mu of fish pond, i.e., the ratio is 1:1. It is also founded that the silkworm excreta contains a higher percentage of nitrogen than pig, sheep, or cattle excreta. Different species of fish at different stages of growth are mixed to maximize the use of pond space and to facilitate harvesting in stages. Newly introduced production eco-system has suggested that one based on sugar cane, animals (fed with sugar-cane leaves and other feed), and the raising of fish fed with animal excreta would reduce the crop disease.

Additional resources, however, are needed to help conduct further research and development to understand how important eco-friendly principles associated with the traditional, sustainable models could be incorporated into modern sericulture production processes at the commercial level. The research team intends to extend this study to where similar practices are still being used so that a comparative analysis could be conducted with the case in the Pearl River Delta to help identify the necessary social, economic, political and institutional measures for reviving, preserving, promoting, and applying LINK-based sustainable resource management in the subtropical regions with high population pressure on the land in other countries.

15

Mulberry: The Worldwide Forage

Mulberry, originated from temperate zones in Asia, has now spread throughout the world. In tropical Central and South America, it has adapted excellently. It is a perennial tree or shrub that is easily propagated, with fast growth and vigorous shooting. It develops a strong vertical and profuse horizontal root system. These features improve physical soil conditions and allow better water conservation. As forage, mulberry has shown excellent organoleptic intake for livestock.

Protein content varies from 14 to 22 percent on a dry matter (DM) basis and *in vitro* digestibility between 70 and 80 percent. Under tropical conditions, where conditions favour plant growth, mulberry culture has been practised under various densities and cutting frequencies, which result in large yield differences. In Guatemala, Blanco obtained 19 tonnes of DM/ha in four cuts every nine weeks, with 30 cm between plants and a cutting height of 75 cm.

In Costa Rica, Benavides, Lachanx and Fuentes used densities of almost 23 000 plants per ha, with DM yields from 21 to 28 tonnes/ha. Rodríguez, Arias and Quiñones worked in Guatemala with planting distances of 60 and 80 cm between plants and harvesting frequencies of six, nine and 12 weeks, with DM yields of 1 to 4.6 tonnes/ha/harvest. These results suggest that yield increases with density. In Japan two planting densities are used, 10 000 and 20 000 plants/ha for the traditional and intensive production systems, respectively.

Mulberry (*Morus* spp.) leaves have long been the traditional feed for the silkworm (*Bombyx mori*). There is evidence that sericulture started about 5 000 years ago and hence the domestication of mulberry. Mulberry has been selected and improved for leaf quality and yield over the centuries. Through silk production projects, mulberry has been taken to countries all over the world, and it has now spread from the temperate areas of northwest and central Asia, Europe and North America through the tropics of Asia, Africa and Latin America to the southern hemisphere (Southern Africa and South America).

There are mulberry varieties for many environments, from sea level to altitudes of 4 000m, and from the humid tropics to semi-arid lands, such as in the Near East with 250 mm of annual rainfall and the southwestern United States. Mulberry is also produced under irrigation. Although the majority of silk production projects have had limited duration because of silk processing constraints and limited market opportunities, mulberry trees have remained in most places where they have been introduced.

The main use of mulberry globally is as feed for the silkworm but, depending on the location, it is also appreciated for its fruit (consumed fresh, in juice or as preserves), as a delicious vegetable (young leaves and stems), for its medicinal properties in infusions (mulberry leaf tea), for landscaping and as animal feed. In Peru the multiple uses of mulberry have been recognised. There are several countries where mulberry is utilised traditionally as a feed in mixed forage diets for ruminants, such as in certain areas of India, China and Afghanistan.

In Italy there have been several studies on the use of mulberry for dairy cows and other domestic animals and in France a research project was undertaken to introduce mulberry in livestock production. But it was only in the 1980s that specific interest in the intensive cultivation and use of mulberry as animal feed started in Latin America. It is surprising that a plant that has been improved for leaf quality and yield to feed the silk worm, which has high nutritional feed requirements, has received such limited attention from livestock producers, technicians and researchers.

Like several significant breakthroughs in science and technology, the discovery of the value of mulberry as a high-quality feed in Latin America happened serendipitously. A Costa Rican farmer of Chinese origin, whose silk project had failed, fed mulberry leaves to his goats and was impressed by the palatability of the leaves and by the performance of his animals. He communicated his observations to scientists at the Tropical Agriculture Research and Training Center (CATIE), who were receptive to the farmer's news and forward thinking enough to include mulberry in their tree fodder evaluations and later in agronomic and animal performance trials.

In Africa, the International Centre for Research in Agroforestry (ICRAF) in Kenya and the Livestock Production Research Institute in the United Republic of Tanzania have also conducted successful agronomic and animal trials by themselves, apparently without being aware of the interest elsewhere.

Mulberry belongs to the Moraceae family (subtype angiosperms; class dicotyledons; subclass urticales) and there are several species: *Morus alba, M. nigra, M. indica, M. laevigata, M. bombycis,* etc. which have been used directly, or through crossings and induced mutations, for the development of varieties to support silkworm production. The diploid *M. alba* (2n=2x=28) is the species most widely spread, but polyploid varieties, which originated in various research stations in Asia, show greater leaf yields and quality.

In general, polyploid varieties have thicker and larger leaves of a darker green colour, and produce more leaves. There is a large variation in leaf production and in leaf quality (e.g. protein content) among the many species and in the varieties and cultivars grown at different locations under a wide range of soil and environmental conditions, indicating the huge potential for identifying suitable germoplasm for most sites. Many of the references on mulberry in the literature do not specify which species or varieties were used. Names are often given based on leaf features. In many cases, locally grown varieties (native or criolla) seem to perform adequately, since they are probably well adapted to local conditions.

Composition of Mulberry

The results of chemical composition of mulberry fractions from various authors are given in Table 1. Crude protein content in leaves varies from as low as 15 percent to 28 percent depending on the variety, age of the leaves and growing conditions. In general, crude protein values can be considered similar to most legume forages. Fibre fractions are low in mulberry leaves compared with other foliages. Shayo reported lignin (acid detergent lignin) contents of 8.1 percent and 7.1 percent for leaves and bark, respectively.

Table 1. Chemical composition (percentaje of dry matter) of mulberry

Variety	CP	CF	NDF	ADF	EE	Ash	Ca	P
Leaf								
Hebba	15.9	12.6	7.1	15.9	2.42	0.24		
Izatnagar	15.0	15.3	7.4	14.3	2.41	0.24		
Palampur	15.0	11.8			5.1	15.5		
Parbhani	22.1	5.9			3.9	13.4	3.3	1.43
Kanva-2	16.7	11.3	32.3		3.0	17.3	1.80	0.14
Mpwapwa	18.6		24.6	20.8		14.3		
Dominican	20.0			23.1	4.0	4.5	2.70	
Criolla	19.8						1.90	0.28
Tigreada	21.1						2.74	0.38
Indonesia	20.1						2.87	0.33
Leaf & young stem								
Tigreada	27.6	13.2				10.4		0.20
Indonesia	24.3	15.3				11.2		0.29
Criolla	27.6	16.9				11.8		0.26
Acorazonada	25.2	14.1				13.4		0.15
Koruso 21	11.0	10.0	22.0	20.6	5.9	13.9	3.13	0.37
Koruso 21	8.0	11.8	24.7	24.5	5.3	19.3	4.76	0.37
Young stem								
Criolla	11.3						1.33	0.29
Tigreada	11.7						1.38	0.33
Indonesia	11.9						1.53	0.43
Dominican	4.7			48.2	1.7	1.3	1.61	
Stem								
Dominican	3.8			50.2	1.0	1.8	1.10	
Mallur	11.5	34.0			2.7	9.32	1.56	0.20
Bark								
Mpwapwa	7.8	46.8	36.9		6.1			
Whole plant								
Dominican	11.3			34.4	1.6	1.9	2.10	

A striking feature of mulberry leaves is the mineral content, with ash values up to 25 percent. Typical calcium contents are around 1.8 to 2.4 percent and phosphorus 0.14 to 0.24 percent. Espinoza, Benavides and Ferreire, found potassium values of 1.90-2.87 percent in leaves and 1.33-1.53 percent in young stems, and magnesium contents of 0.47-0.63 percent for leaves and 0.26-0.35 percent for young stems. Table 2 shows the digestibility of mulberry.

Table 2. Digestibility of mulberry

Method	*Fraction*	*Digestibility (%)*
In vivo (goats)	Leaf	78.4-80.8
In vitro	Leaf	89.2
	Leaf	80.2
	Leaf	89-95
	Stem	37-44
	Total	58-79
	Leaf	82.1
	Bark	60.3

As can be seen, leaf digestibilities *in vivo* (goats) and *in vitro* are very high (>80 percent) and total digestibility is equivalent to that of most tropical forages. The degradation characteristics of mulberry, determined by the nylon bag technique, are presented in Table 3. Leaves would be completely degraded if they remained in the rumen for enough time.

The average amino acid composition and N content of 119 mulberry varieties grown experimentally in Japan are shown in Table 4. Tryptophane was not included in the analysis. As can be seen from the data, essential amino acids are over 46 percent of total amino acids. It can be calculated from the table that the average nitrogen (N) is 16.6 percent of the total molecular weight of the mulberry amino acids (plus ammonia), and thus the converting factor from N to mulberry protein is 6.02. The 204.3 mg of amino acids per g of protein is equivalent to 3.47 percent N, which is 80 percent of total N in mulberry leaves.

Table 3. In sacco degradation of mulberry

Fraction	Parameter			
	a	*b*	*a + b*	*c*
Leaf	35.7	64.0	99.7	0.0621
Whole plant	30.4	46.2	·76.6	0.0667
Leaf & young stem	27.8	48.95	76.8	0.0300

Once tryptophane is subtracted, the difference, a non-protein fraction, is likely to be composed of nucleic acids and other unidentified N compounds. The most important protein in mulberry leaves, as in most leaves, is ribulose-1.5-bisphosphate carboxylase (RuBisCO) whose active site is responsible for carbon fixation. Nitrogen in RuBisCO can be 43 percent of the total nitrogen in mulberry.

Palatability: One of the main features of mulberry as forage is its high palatability. Small ruminants avidly consume the fresh leaves and the young stems first, even if they have never been exposed to it before. Then, if the branches are offered unchopped, they may tear them off and eat the bark. Cattle consume the whole biomass if it is finely chopped. There is a report of ad libitum DM intake of 4.18 percent of liveweight (average of three lactating goats), which is much higher than in other tree fodders. Jayal and Kehar reported DM intakes of mulberry leaves of 3.44 percent of body weight in sheep under experimental conditions.

Table 4. Amino acids and N content of mulberry varieties and soybean meal

Compound	Soybean meal		Mulberry		
	content (mg/g DM)	%1	Content (mg/g DM)	SD	%1
Non essential amino acids	n.a.		108.93	53.3	
Essential amino acids (EAA):					
Lysine	32.92	6.7	12.33	2.58	6.0
Methionine	7.30	1.5	2.99	0.61	1.5
Threonine	20.34	4.1	10.52	1.75	5.2
Valine	26.29	5.3	12.83	2.17	6.3
Isoleucine	26.85	5.4	10.04	1.88	4.9
Leucine	39.55	8.0	19.45	3.10	3.1
Tyrosine	14.38	2.9	7.40	1.39	3.6
Phenylalanine	25.51	5.2	12.26	2.06	6.0
Histidine	12.92	2.6	4.61	0.82	2.3
Trytophane	6.97	1.4	na	-	-
Total EAA	213.03	43.1	92.43	-	45.3
Ammonia (NH_3)	na		2.89	0.54	1.4
Total (AA + NH_3)	494.38	100	204.25		100
Nitrogen (percent)	7.91		4.36	9.63	

Animals initially prefer mulberry over other forages when they are offered simultaneously, and will even dig through a pile of various forages to look for it. In a comparative study, Prasad and Reddy reported higher daily dry matter intakes of mulberry leaves in sheep than in goats (3.55 against 2.74 kg DM/100 kg body weight).

Agronomy

Establishment: The most common planting method worldwide is by stem cuttings, but in certain places seed is preferred. As is the case with other tropical perennial forages for cut-and-carry systems, planting by seed assures deeper roots with a greater capacity to find water and nutrients that eventually results in higher biomass production and greater longevity. Seeds might be the most acceptable way of transporting, quarantine and store selected materials. The advantages of stem reproduction (cloning) are certainty of production characteristics, practicality in obtaining material and ease of planting.

Male plants might be preferred when introducing foreign germoplasm to new locations since this prevents involuntary expansion. As in most perennial forages, the time and the establishment cost (mainly for land preparation, planting and weed control) are critical for the successful introduction of mulberry.

Cultivation: Mulberry is cultivated for fruit as isolated trees or in orchards; for small-scale silkworm rearing along boundaries or along food crops in mixed farming systems; for large silk projects or for intensive forage production in pure stands; and also for forage in association with N-fixing legumes. Mulberry is also found mixed with other trees in natural forests or plantations.

Fertilisation: All the required nutrients for mulberry growth must come from the soil, since mulberry does not fix atmospheric nitrogen. In pure stands, mineral and organic fertilisers (animal and vegetable manures) must be used to replenish the nutrients removed with the foliage in order to maintain a sustainable production. The association with legumes with effective N-fixing rhizobium can reduce N inputs and may be the most desirable combination for some farms, but even when recycling nutrients in animal manures, extra chemical fertilisers are required for maximum yields.

Responses of mulberry to N fertilisers have been clearly demonstrated, both in inorganic and organic forms, with better responses to the latter (Table 5). According to Kamimura et al., the nitrogen level in soils is the major factor for mulberry growth.

Table 5. Effect of goat manure or ammonium nitrate application on total DM yields during three consecutive years

Year	Level of manure (tonnes DM/ha/year)				NH_4NO_3
	0	240	360	480	480
1	23.0	24.4	26.6	31.1	26.7
2	21.3	25.2	27.6	33.4	29.7
3	22.9	28.2	32.6	38.2	29.2

Harvest and preservation: For silkworm feeding, individual leaf picking, shoot harvesting and whole branch cutting are practised, depending on the feed requirements of the silkworm larvae stages and harvesting costs. For silkworms, leaves are offered fresh some other forms of feeding are being developed. For ruminant feeding, the preferred method has been branch cutting by hand, although mechanical harvesting could be employed in the future for direct feeding of fresh material on a large scale, for processing or for drying.

Forage conservation by ensiling has been successfully achieved and there have been some preliminary studies on leaf drying. Leaf blades dry within hours under full sun but more time is required for petioles and stems. Some conditioning may help to reduce water content and minimise the deterioration of leaf quality by over exposure. Diploid varieties dry more quickly since they tend to have more stomata per unit of leaf area.

Yields: The production of leaf and total DM per hectare of mulberry depends on the variety, the location, plant density, fertiliser applications and harvesting techniques. Table 6 gives the yields of mulberry in various locations.

Table 6. Examples of mulberry yields

Location	*Variety*	*Fraction*	*Yield (tonnes/ha/year)* Fresh	DM
Karnataka, India	M-5	Leaf	40	
		Stem	52	
Mpwapwa, Tanzania	Local	Leaf		8.5
		Stem		14.1
		Bark		2.7
San José, Costa Rica	Tigreada	Leaf and young stem		11.0
	Indonesia	Leaf and young stem		8.7
Puntarenas, Costa Rica	Tigreada	Leaf and young stem		13.4
	Indonesia	Leaf and young stem		12.5
Matanzas, Cuba	Tigreada	Total biomass	30	
	Acorazonada	Total biomass	33	
	Indonesia	Total biomass	26	
	Local	Total biomass	30	
Cuyutla, Guatemala	Local	Total biomass	37	
		Leaves	16	
Zhenjiang, Jiangsu, China	Shin Ichinose	Leaves	32	
		Branches	28	
		Stems	8	
Kalimpong, W. Bengal, India	Local	Leaves	22	
	BC 259	Leaves	20	
	TR 10	Leaves	19	
	C 763	Leaves	19	

Total biomass yield and leaf proportion vary with species and varieties. Climate (moisture and solar radiation) and soil fertility are determining factors on productivity. Increasing planting density increases leaf yields. Fresh leaf yields of 40 tonnes/ha/year (approximately 10 tonnes of dry matter) have been reported in India and in Costa Rica. Maximum dry matter yields of edible material (leaves and young stems) and total biomass were 15.5 and 45.2 tonnes/ha/year, respectively. Total leaf DM yields of less than 10 tonnes could be expected under less intensive production.

Animal Performance with Mulberry

Ruminants

Although the feeding value of mulberry for dairy cattle has been recognised for some time in Italy and it has been traditionally used in Himalayan countries, research on mulberry for ruminants has been rather limited. Jayal and Kehar, given on the high digestibility values of *M. indica* leaves, suggested that they could be used as supplements for lower-quality forages. Mulberry was used to replace grain-based concentrates in lactating cows with excellent results (Table 7). Yields did not significantly decrease when 75 percent of the concentrate was replaced with mulberry.

Milk production increased with the levels of mulberry offered to goats on a King grass diet as shown in Figure 1. At CATIE, Turrialba, Costa Rica, a module of two dairy goats (Saanen x Toggenburg), fed exclusively with forage from 775 m^2 of mulberry (17 000 plants/ha), in association with *Erythrina berteroana* (5 128 trees/ha) just as green manure, and from 425 m^2 of King grass, produced an average of four litres per day, equivalent to over 12 000 litres per ha/year.

Table 7. Substitution of concentrates by mulberry in lactating Holstein cows grazing Kikuyu grass (Pennisetum clandestinum)

Parameter	Concentrate: Mulberry 100:0	60:40	25:75
Milk yield (kg/d)	14.2	13.2	13.8
Intake (kg MS/d):			
Concentrate	6.4	4.2	1.9
Mulberry	0	2.8	5.5
Kikuyo grass	9.3	7.8	6.2
Total	15.7	14.8	13.6

Also in Costa Rica, liveweight gains of bulls belonging to the Romosinuano breed (a criollo type) fed Elephant grass, increased to over 900 g/day when mulberry was offered as a supplement at 1.7 percent of their body weight on a DM basis. Table 8 presents the results of an experiment in Guatemala with growing Zebu x Brown Swiss steers being fed increasing levels of mulberry as a supplement to a sorghum silage diet. Although the growing rates with the highest mulberry level are not impressive (195 g/day), most likely because of the poor quality of the silage, this trial shows the high nutritive value of the supplement.

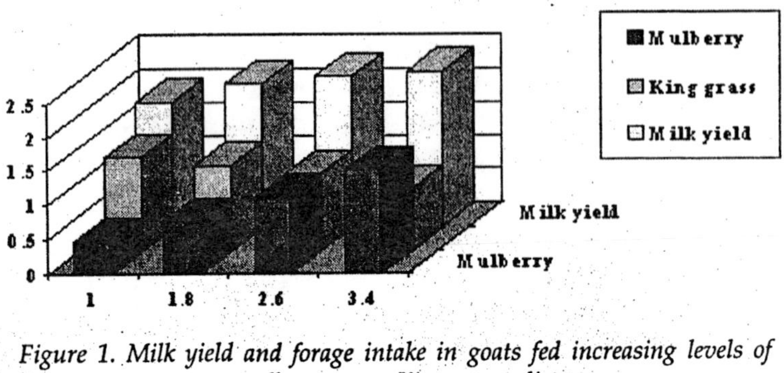

Figure 1. Milk yield and forage intake in goats fed increasing levels of mulberry on a King grass diet

Total intake and weight changes improved with the amount of mulberry offered reflecting its higher nutritive value compared to the basal diet. Daily gains of female calves (0-4 months) were not affected when mulberry leaves were offered ad libitum and the commercial concentrate reduced to 25 percent of the amount traditionally used. In lambs, gains reached 100 g/day when King grass was supplemented with 1.5 percent dry matter of mulberry.

Monogastrics

The silkworm has a relatively simple digestive system. In some ways it is comparable to that of monogastric animals, so that, in theory, mulberry leaves could also be used at least as one of the ingredients in monogastric diets. In a trial with growing pigs in which a commercial concentrate was replaced by up to 20 percent of mulberry leaf, the best level of substitution was 15 percent. It increased daily gains from 680 g/day ith only concentrates, to 740 g/day and also gave the best economic results. In rabbits, the reduction of concentrate offered daily from 110 to 17.5 g with ad libitum fresh mulberry only reduced gains from 24 to 18 g/day, but decreased to more than half the cost of the meat produced.

The combination of mulberry and *Trichantera gigantea* leaves, as the protein source, and blocks made of molasses, cassava root meal and rice bran, as the energy source, gave better reproduction and growth performance than a diet of commercial concentrates and grass. Singh, Goel and Negi supplemented Angora rabbits, receiving pelleted diets, with mulberry leaves ad libitum and obtained intakes of mulberry equivalent to 29-38 percent of the total intake. This level significantly reduces feed cost.

Deshmukh, Pathak and Takalikar fed mulberry leaves as the sole ration for adult rabbits. These authors found daily intakes of 68.5 g for dry matter, 11.2 g for crude protein and 175 kcal for digestible energy (equivalent to 2.55 Mcal of digestible energy per kg). The digestibility values were 74 percent for crude protein, 59 percent for crude fiber and 64 percent for dry matter. The authors concluded that mulberry leaves provided enough nutrients for maintenance. Narayana and Setty found better egg yolk colour and increased egg size and production with the inclusion (up to 6 percent) of shade-dried *M. indica* leaf meal in the mash of laying hens.

Table 8. Effect of mulberry supplementation level on intake and weight changes of Zebu x Brown Swiss steers fed on sorghum silage

Parameter	Mulberry level (% BW)			
	0	0.5	1.0	1.5
DM intake (percent BW/day				
Total	2.26	2.39	2.64	2.88
Sorghum silage	2.26	1.91	1.68	1.51
Daily gains (g/d)	- 128	- 29	164	195

Other small herbivores, such as guinea pigs, iguanas and snails, could also be fed mulberry leaves. In fact, wild green iguanas (*Iguana iguana*) came to feed on recently established mulberry fields in Costa Rica.

Livestock Production Systems

The traditional way of using mulberry as animal feed in silkproducing areas is by providing ruminants with the residue left by the silkworm. A model of sericulture and milk production has been proposed by Mehla, Patel and Tripathi, in which dairy cows receive mulberry residue and concentrates. The generation of edible protein and employment are much greater than with food grains. This refuse material is added to fish ponds for herbivorous carps in the Chinese dyke-pond system, which is one of the most intensive agricultural low-input systems in the world, and generates food and outputs for a large number of people.

In these silk areas, as well as where mulberry grows wild, cut-and-carry systems are practised and it is the most obvious way of utilising mulberry for livestock, either from pure stands or from associations with legumes. Mulberry foliage can constitute the supplement to low-quality forage (grass) based diets or as the main component of the ration. A natural association of mulberry and livestock occurs in regions such as the Near East and Central Asia where mulberry trees are kept for fruit production. Fallen leaves in the autumn are consumed by domestic animals. Since fruit ripens in late spring or early summer, it may be possible to harvest leaves for forage one or more times before the winter.

The only suggestion of utilising mulberry for direct grazing came from FAO, proposing a complementary association with clover (*Trifolium subterraneum*) for sheep and cattle grazing in Tuscany (Italy). Mulberry benefits from the N fixation by the clover and contributes with high-quality forage during the summer. The association produces more forage over a longer period than the individual pure stands.

Bibliography

Chinnaswamy, K.P. & Hari Prasad, K.B., "Fuel energy potentiality of mulberry", *Indian Silk*, 1995.

Choudhury, P.C., "Mulberry cultivation", CSRTI, Mysore, *Cultivation technology for chawki mulberry garden.*

_______________., "Handbook of silkworm rearing", *Agriculture Technique Manual* 1, CSRTI, Mysore, Tokyo, Japan, Fuji Publishing Co. Ltd, 1972.

_______________., "Chemical composition and quality of mulberry leaves", *Text book of tropical sericulture*, 1976.

Jalaja S. Kumar, Sarkar, A. & Datta, R.K., "A breakthrough in mulberry breeding in sustainable cocoon production". *In Global silk scenario - 2001*, Proceedings of the *International Conference on Sericulture*, p.242-247. Oxford and IBH publishing Co. Pvt. Ltd,

Rangaswami, G. Narasimhanna, M.N., Kasi Viswanathan K. & Sastry, C.R., *Manual on Sericulture Vol.1*, Rome, FAO, 1976.

Ravindran, S., Ananda Rao, A., Girish Naik, V., Tikader, A., Mukherjee, P. & Thangavelu, K., "Distribution and variation in mulberry germplasm", *Indian J. Plant Genet. Res.* 1997.

_______________., Tikader, A., Girish Naik, V., Ananda Rao, A. & Mukherjee, P., "Distribution of mulberry species in India and its utilisation", *Poster paper presented in National Dialogue*, 1988.

Sarkar, A. & Jalaja S. Kumar, "Gradual improvement of mulberry varieties under irrigated conditions in South India and potential programme for varietal selection", *International Symposium on Sericulture Science*, Hangzhou, China, 1996.

Susheelamma, B.N., Jolly, M.S., Sengupta, K., Giridhar, K., Baksh, S., Mogili, T., Mallikarjuna, R.S. & Pavan Kumar, T., "Statistical analysis of adaptability of drought resistant mulberry genotypes", *Sericologia,* 1992

Sastry, C.R., "Mulberry varieties, exploitation and pathology", *Sericologia,* 1984.

Ullal, S.R. & Narasimhanna, M.N., *Handbook of Practical Sericulture,* Bangalore, Central Silk Board, 1987.

Watt, G., *Dictionary of economic products of India,* Vol. 5, p. 279-284. Delhi, Periodical Experts, 1981.

Index